Science, Mysticism and Psychical Research

Science, Mysticism and Psychical Research:

The Revolutionary Synthesis of Michael Whiteman

By

John Poynton

Cambridge Scholars Publishing

Science, Mysticism and Psychical Research:
The Revolutionary Synthesis of Michael Whiteman

By John Poynton

This book first published 2015

Cambridge Scholars Publishing

Lady Stephenson Library, Newcastle upon Tyne, NE6 2PA, UK

British Library Cataloguing in Publication Data
A catalogue record for this book is available from the British Library

ISBN (10): 1-4438-8019-1
ISBN (13): 978-1-4438-8019-0

For Sibyl

Contents

List of Figures and Tables xi

Foreword: Michael Whiteman, A Unique Individual xii
Prof. Vernon M Neppe

Preface xvi

Acknowledgements xxi

Part I: Basics of Whiteman's Thinking

Chapter One 2
A Discordant Trio?

Chapter Two 7
Potentiality and Actualisation

Chapter Three 11
Physical and Non-Physical

Chapter Four 15
Classifying Non-Physical States

Chapter Five 19
Separative Experience

Chapter Six 24
Reality of Separative Experience

Chapter Seven 29
Other Spaces

Chapter Eight 33
Scientific Thinking and Practice

Chapter Nine 37
Space, Time and Release

Chapter Ten 42
Physics and Psychology

Chapter Eleven 50
The Iron Rule and Immanuel Kant

Part II: The House of Personality

Chapter Twelve 56
The Mind of a Murderer

Chapter Thirteen 61
The Corporate Structure of Personality

Chapter Fourteen 66
Reincarnation and Alternatives

Chapter Fifteen 70
Tangled Lines of Consciousness

Chapter Sixteen 75
Inner Contests and Skills

Chapter Seventeen 82
Sexuality and Chakras

Chapter Eighteen 88
Crime and Punishment

Chapter Nineteen 94
Destiny and Release

Part III: Spiritual Development

Chapter Twenty 100
Spirit, Letter, and Mystical Experience

Chapter Twenty-One 104
The Yoga Sutras

Chapter Twenty-Two 112
Before the Yoga Sutras

Chapter Twenty-Three 116
The Minoan Culture

Chapter Twenty-Four 120
The "People": Isaiah and the Psalms

Chapter Twenty-Five 125
The "Son": St Paul and St John

Part IV: Elements of Mystical Experience

Chapter Twenty-Six 132
Life in Other Worlds

Chapter Twenty-Seven 135
Lower and Higher Worlds

Chapter Twenty-Eight 139
The Archetypal

Chapter Twenty-Nine 142
Deletion and Transformation

Part V: Overview

Chapter Thirty 148
"The Meaning of Life"

Chapter Thirty-One 154
A Mystical Life

Chapter Thirty-Two 159
Substructure: Bed for a Harmonious Trio

Appendix A 163
The Mystical Derivation of Physical Laws

Appendix B.. 175
A Brief Biography

References 178

Index.. 182

LIST OF FIGURES AND TABLES

Table 6-1 General Index of Reality 26

Table 10-1 Objective and Subjective Cycles 47

Figure 23-1 Fresco from the Palace of Knossos 117

Figure 23-2 Minoan Depiction of Functions 118

Figure Appendix B-1 Michael Whiteman Aged 84 175

FOREWORD: MICHAEL WHITEMAN, A UNIQUE INDIVIDUAL

It is an honour for me to write this Foreword to Professor John Poynton's book *Science, Mysticism and Psychical Research: The Revolutionary Synthesis Of Michael Whiteman.*

This is because, besides Professor Poynton, I may be one of the few to have had significant contact with Professor Whiteman in the discipline of this book's focus, namely Psychical Research, or research into consciousness. This I was able to do, firstly, during the early 1980s as erstwhile President of the South African Society for Psychical Research (SASPR) and Editor of the *Parapsychological Journal of South Africa (PJSA).* Secondly, after relocating to the United States in 1986, I would continue to visit Dr. Whiteman at his Cape Town home, and interchange ideas.

I can write objective information about the genius of Michael Whiteman, but John Poynton has documented that beautifully in this remarkable book. So, for example, Prof. Whiteman was the recipient of the Marius Valkhoff award for "exceptional contributions to psychical research"; this was a South African international award given occasionally (eleven recipients in 30 years).

Instead, I approach this Foreword more from a personal perspective.

I would be asked, who is this Michael Whiteman you're meeting with? I would usually reply:

"Michael represents the four Ms: Michael the mystic, mathematician, and musician. He is, in my opinion, the greatest genius living on the continent of Africa. His work is so esoteric few understand it, and I have the great honour of carrying through his great legacy for future generations."

That term "future" is important here. His contributions will endure like some of the great philosophers for centuries.

I remember him submitting a very lengthy esoteric article on his spiritual philosophy to me while I was editor of the *PJSA.* The referees had written that they "never understood a word, but knew it was critical to put into print". I recognized this specialness as editor. But there was a problem: I wrote to Michael in Cape Town (I was in Johannesburg; and those were the days of long-hand writing) and indicated that we very much wanted to publish it but could he abbreviate it? His reply was typical. He wrote back and said every word was critical and he could not. So I published the article in toto and it took up half the journal. I knew it was not for people then, but his contributions would long endure into a time when mystics could better understand his world.

I remember spending what must have been three continuous hours in animated esoteric and complex discourse, inter alia, about dimensions of space and time. A cousin, with whom I was staying, had dropped me off, and sat all that time with us, never saying a word. After we left, my cousin said to me: "That was the most amazing intellectual experience of my life. I hardly understood a word but the concepts will remain. I realize I was witnessing a unique, great interchange that will stay with me forever."

This was Michael, a great man, who was a giant amongst giants.

I was always intrigued by Michael's theoretical and practical awarenesses of Out of Body Experience—what he preferred to call Separative Experience (SE)—and as per his wont, had phenomenologically classified SE into subtypes as Dr Poynton has pointed out. By the time I visited him two and a quarter years before he passed, he had had over ten thousand deliberately induced separative experiences. Like a true scientist each had been carefully documented—a massive contribution because no-one ever has had that amount of documented SEs.

I looked at this ostensibly middle-aged looking man who lived on his own in a two story home and would move with facility between floors and take a brisk walk for several hours per day before settling down to his avocation of writing. If I had not known differently, I would have regarded him as someone in his early seventies, but I knew it could not be. *Perhaps in his 80s but he must be older than that*, I thought.

"Michael, how old are you, exactly?" I asked.

"Next month, I will be 98", he replied.

I quipped there was a reason for his longevity: Time out of the body did not count!

He described to me his very regimented existence even at age 97. It went something like this:

Get up early, eat, do some reading for 2 hours, walking for 2 hours, meditating for 2 hours, writing for 2 hours, resting for 2 hours. His meditation often involved his deliberate out of body experiences, and in this he may have been assuming a different kind of experiential identity, neither male nor female or it was both, or reflecting the essence of one of the sexes because he had a great interest in essence of self, in different kinds of souls, and like I did, felt that we were part of a complex multidimensional existence.

One of the few times Michael made the news was apparently in a lengthy article in the local newspaper about this Mathematics Professor who turned a hundred. I did not see it but was told there was no parapsychological mentions and the layperson may not have realized what the titles of his books implied. As I recall, his amazing language versatility e.g. Sanskrit and many different related dialects through to Hebrew, may have been mentioned in that article.

In February 2007, on a group called SurvivalNet, we wrote about Michael for the last time in a thread started by someone else. Discussing Michael on this group was rare, and I had written about Michael last on 23 June 2003. My response content at that time (June 2003) is worth sharing briefly because it clarifies Michael and his role in the world a little more.

> … You are referring to Professor J H Michael Whiteman, mathematician, musician and mystic. He was a professor of Applied Mathematics at the University of Cape Town. He was also a theoretical physicist, philosopher and metaphysician. A remarkable genius, Michael had developed the OBE state to a profound degree. He could actively, subjectively project himself out of his body at will. He developed the term "separative experience" (SE) for OBE, and described several levels of SE. His theoretical contributions are unfortunately seldom cited by parapsychologists today… He wrote several books. I have hard copies of some of his articles. … Michael was certainly known amongst those that followed his work as a remarkable mystic but knowledge about his work outside South Africa seems to have been limited.
>
> I hypothesize his lack of attention was threefold:
>
> 1. Lack of adequate publicist and his philosophy not of itself being newsworthy.
>
> 2. South Africa was not a fertile area for one's research or theory to be publicised elsewhere. He did not tend to publish much outside South Africa other than e.g. the *Journal of the Society for Psychical Research* at times. … He contributed on several occasions to the *Parapsychological J. of South Africa*, which was abstracted by Psych Info during its tenure of existence, and one of these was a lengthy letter on OBEs. He also published in several issues of a journal called *Parapsychologica*, as well as John Poynton's book *Parapsychology in South Africa* in 1975 (at which point he was already Emeritus Faculty). Of course the Internet now has changed this, and we are all now part of this Global Internet Village.
>
> 3. His work was so complex it was understood by very few.

What John Poynton has done in this book is make the major unintelligible contribution of Michael Whiteman, intelligible. This is enormously important not only now, but for the future as Whiteman was a man for the ages. Here is Dr. Whiteman, a giant for all time, who died forgotten because others did not understand his contributions. And Dr Poynton has ensured that his 181 page book in five parts, thirty two chapters and two appendices can be understood, and interpreted for sheer mortals in a comprehensible manner. Moreover, there is no-one in the world who could do it better than John, having spent a very large amount of time studying Michael's contributions. Therefore, Poynton's *Science, Mysticism And Psychical Research* is a major and important work emphasizing the work of a great pioneer of Psychical Research and Mysticism.

Prof. Vernon M Neppe MD, PhD, Fellow Royal Society (SAf), BN&NP, DFAPA, MMed (Psych), DPM
Director, Pacific Neuropsychiatric Institute (www.pni.org), Seattle, WA, USA
The Vernon Neppe Gateway (www.VernonNeppe.org):
Vernon Neppe Research Site (VernonNeppe.com)
Executive Director and Distinguished Professor, The Exceptional Creative Achievement Organization (www.ecao.us)
Director, Human Performance Enhancement Division (HPED) of the World Institute of Scientific Exploration (WISE).

PREFACE

Many people see this century as bringing in a new Renaissance, the rediscovery and growth of what is now commonly called "spirituality". This exploration of experience and states beyond physical sensing is pitched against an entrenched materialism (or matterism), which may be seen as an impoverishment of life and living, rife with false perceptions. The new renaissance aims at a "reconciliation of science and spirituality", as portrayed by Kelly *et al.* in *Beyond Physicalism* (2015). It recognises "irreducible mind" (Kelly *et al.*, 2007) and has been called post-materialist science (Beauregard *et al.*, 2014). Yet it could easily go off the rails into a bog of its own false perceptions, so there is need for rigorous observation and thinking.

I hope even the first chapter of this book will suggest that rigour is available in the work of Michael Whiteman, late associate professor of applied mathematics at the University of Cape Town, South Africa. The rigour in his work had a revolutionary purpose, melding mathematical physics and general science with Indian and Western mystical texts, clarified by a life-time of psychical and mystical experience, and coupled with an extensive knowledge of philosophy, psychology, and psychical research or parapsychology. His work is mind-stretching to the point that led one of my most able ex-students to write that every time she read Whiteman her brain hurt. I know the feeling. I would like to lessen the pain by laying a negotiable pathway through landscape that might seem impenetrably complex and unfamiliar.

Previously I have written on the relationship between Whiteman's ideas and psychical research (Poynton 1994, 2001, 2011), to which several people responded by suggesting that I should write a larger study. But there are several difficulties in doing so. From the writer's point of view one has to sift through Whiteman's rather dense literary style to reveal his nuggets of wisdom. His writing guarded against ambiguity and criticism to the point of laborious reading. Also, despite painstaking revisions, he tended to race ahead of his readers and not introduce or express his complex ideas too clearly. The choice of a compiler of his work is either to pick out and copy slices of text, or to transcribe the text into his/her own words and understanding. I have chosen the former option, with a fair amount of the latter. Where he is clear he is very clear, and often he has a neat turn of phrase that is worth preserving.

Some editing is nevertheless required. The text is cluttered with quotation marks, which he inserted when a word is used in a technical mystical sense; these are deleted when they seem unnecessary. When technical words of this kind were used frequently he italicised them, but this again impairs reading and they are converted to Roman when italics do not seem needed. A few of his readers have found his inclusion of Pali, Sanskrit, Greek and Hebrew words distracting, breaking up the continuity of reading. Pali words are close to the language mainly used by Gotama Buddha, Whiteman's primary reference. He treated these as technical terms essential to mystical thinking, and they are mostly retained in textual extracts; words from other languages are usually deleted from passages in the interests of easier reading, unless thought to be of special importance. Hopefully this editing will give a smoother ride through his texts, and hurt the brain less.

Then there is the difficulty of parcelling his ideas coherently. His range of experience, insight, scholarship and understanding extended well beyond a usual range of experience and thought, so most readers will have only a relatively limited personal context in which to place the work. This makes it hard to grasp his meaning. Perhaps the only qualification I have for attempting to cover his thinking is that for the last thirty-five years of his life he treated me as his guinea-pig, as he often called it, for trying out drafts of his work. I am a biologist with academic background in philosophy and music, and long association with psychical research, but only minor academic acquaintance with most subjects covered in his work. Perhaps life as a biologist makes you more ready to deal with anything thrown your way, which is how Whiteman seemed to regard me.

Outside his personal correspondence, he did not comment at length on the work of other contemporary writers on mysticism and philosophy of science, nor did he engage in general reviews such as the chapters on mystical experience by Kelly (2007) and Marshall (2015). Virtually the only point he repeatedly made about the existing literature on mysticism was that authors rarely show signs of extensive personal mystical experience, and so are liable to misunderstand the basics of authentic reports. Consequently this book does not, to any length, compare and contrast his thinking with that of writers on mysticism in modern times, apart from the influential philosopher Immanuel Kant. The attention Whiteman gave to the scientist-turned-mystic, Emanuel Swedenborg, also needs exploring, not least because Swedenborg became the subject of Kant's confused criticism.

In writing about Whiteman's experience and ideas, one has to be aware of longstanding hostility in science and academia towards psychical and mystical research. It is prone to be labelled "pseudoscience", often by people who have no experience as research scientists themselves. Whiteman regarded the hostility to be based on outdated science and poor scholarship, combined with "cognitive pathology" discussed by Maslow (1966), who saw scientists out of their depth as "desperately and stubbornly hanging on to a generalization, in spite of new information that contradicts it." Whiteman's line of thinking was up against the pervading "iron rule of the mechanistic régime", as his great countryman Jan Smuts called it in his seminal *Holism and Evolution* (1926). Therefore an enquiry into the philosophical status of this hostile fashion and its origin is required if science, mysticism and psychical research are to be seen as complementary rather than contradictory. This will be left to a concluding chapter of this book's first Part.

It is also said that psychic phenomena "undermine" science. Whiteman constantly made the point that only pre-twentieth century "classical" science is incompatible with psychic and mystic phenomena. The subsequent revolutionary scientific world-view, to which he himself contributed, provides a unified basis for psychical, spiritual and physical understanding, as this book aims to show. The conventional opposing stance gives the impression of unwillingness to enquire open-mindedly and thoroughly into the methods and data of current physics, psychical and mystical research, and of people's raw experience. Whiteman's ideal for science in all these areas was for it to be "open-minded, rigorously tested, rationally coherent, and illuminating." There is rapidly growing criticism of the conventional hostile attitude, strongly shown in the developing post-materialist "paradigm for science, spirituality, and society" (Beauregard *et al.*, 2014). This study of Whiteman is intended to add to that development in the spirit of Smuts, who advocated an open-minded study of "the psychical or spiritual order".

I have written four novels, but I would never attempt to portray Michael Whiteman. Behind the affable and generous male persona lay depths that were the well-spring of his genius. I have tried to draw what is fairly easily digested from these depths. The first eleven chapters, grouped as Basics of Whiteman's Thinking, are designed to give some insight into core ideas and method. They develop a conceptual framework for mysticism and for psychical research, the latter often taken to task for lacking such a framework. There follow his explorations into personality structure and psychology, sexuality, survival of death and reincarnation, and, finally, Eastern and Western mysticism, spirituality and religion in a broad sense. After a brief account of his own spiritual development there is a return to the relationship between science, mysticism and psychical research.

Even though pared down, the coverage attempted in this book might feel tough going; but a true science of mysticism and psi should never seem an easy ride. Yet Whiteman recognised that his "mystical derivation of quantum theory and physical laws in general" lay well beyond most readers; it was relegated to a supplement in his last book (2006), published a few months before he died. The mathematics of the derivation are accordingly placed in an appendix in this book. Also included as an appendix is a brief biographical sketch. The stages of his mystical development are covered in Chapter 31, but a full "life and works", which would have to include the rich musical career that ran alongside his scientific career, is not attempted here.

Literature citations in this book consist mostly of selected references to Whiteman's publications. To avoid cluttering the text, it seems useful to code them as superscripts rather than give full references each time The publication date followed by the page number at the beginning of a passage seems the most handy code, omitting the century, so that 1986a page 40 becomes [86a,40]; 2006 page 22 becomes [06,22]. For quick reference the codes are set out below. Hopefully the reader will soon associate date with publication; most of them refer to his last four books. Several papers published in scientific journals were reprinted or rewritten in the last two volumes of *Old and New Evidence on the Meaning of Life,* 2000 and 2006. References to both are usually included. Full references are given at the end of the book.

61 *The Mystical Life*
67 *Philosophy of Space and Time*
73 Quantum theory and parapsychology
75a Parapsychology as an analytic-deductive science
75b The scientific evaluation of out-of-the-body experience
75c A three-tier ontology for parapsychology and modern physics
77a Parapsychology and physics
77b The convergence of physics and psychology
79 An introduction to the mystical model for psychopathology
85 Scientific mysticism and altered states
86a *An Introduction to Scientific Mysticism*
86b The mystical way and habitualization of mystical states
93 *Aphorisms on Spiritual Method; the "Yoga Sutras of Patanjali"*
94 Cognitive pathologies and the new paradigm in physics
00 *The Dynamics of Spiritual Development*

03a The problem of discarnate survival
03b On the scientific validation and reality rating of reports of non-physical experience
06 *Universal Theology and Life in Other Worlds*

Citations from his correspondence carry the date of the letter. All letters and documents pertaining to Professor Whiteman are held in the archives of the Society for Psychical Research, housed in the Cambridge University Library. Enquiries should be directed to the SPR Secretary: secretary@spr.ac.uk.

ACKNOWLEDGEMENTS

I first met Michael Whiteman at a 1966 meeting of the South African Society for Psychical Research in Johannesburg, whose formative role I gratefully acknowledge. There followed an exchange of letters and faxes with Professor Whiteman that fill a 12 cm wide box. In his books he warmly acknowledged the forty years of our fruitful contact; it is in sadness for a lost teacher and friend that I reciprocate.

With deep pleasure I acknowledge the friendship and ready help of his daughter Sibyl. As the holder of copyright to her father's books she has made the style and content of this book possible by granting unreserved permission. She also supplied the portrait of her father. My thanks go also to the publisher of Whiteman's last four books, Colin Smythe, for discussion and advice in the preparation of this book. The production of these four major works are much to his credit.

It is a privilege to thank Stanley Krippner for his recommendation, and for granting permission to quote from *Future Science* (1977), which he co-edited. Likewise I thank Vernon Neppe for his generous Foreword. I also thank Bernard Carr, Edward Kelly and Paul Marshall for their recommendations.

I have had valuable discussions and received help from Caroline Watt, Debra Roberts, Paul Marshall, Edward Kelly, Catherine Osborn, Velta Snikere, David Ellis, and Robert McLuhan.

I am grateful to Prometheus Books for permission to include passages from *Materialism: An Affirmative History and Definition* by R.C. Vitzthum (1995), and to the Swedenborg Foundation Press for permission to quote passages from *Kant on Swedenborg: Dreams of a Spirit-Seer and Other Writings,* edited by Gregory R. Johnson (2002).

Last but by no means least I acknowledge the speedy and professional work carried out by Cambridge Scholars Publishing in the preparation of this book.

PART I

BASICS OF WHITEMAN'S THINKING

CHAPTER ONE

A DISCORDANT TRIO?

In Oscar Wilde's *The Importance of being Earnest* we hear that in married life three is company, two is none. But can science, mysticism, and psychical research bed down as a threesome? Conventional thinking would rate even a pair of them to be appalling bedfellows. Science is supposed to explore the real world around us, while mysticism occupies itself with some other-worldliness that people may talk about but which reveals little beyond proneness to fantasy. In psychical research, ESP merely stands for Error Some Place (Honorton, 1975); the claimed data from psychical research cannot be made to fit the real world of science. And parapsychologists generally shy away from mysticism.

Mysticism might seem the worst misfit, the unwanted remains of medievalism. We are required by materialism to recognise the merits of "the Enlightenment and its rejection of mysticism in favour of reason." (Vitzthum, 1995, p. 9). Even from parapsychology comes a view that "A mystic is a person who tries to bypass the evidence of the senses and reasoning in search of deeper truths—just the opposite of a scientist." (Gebelein, 2013, p. 160). Yet one of the books by Michael Whiteman has the subtitle, *An Introduction to Scientific Mysticism.*[86a] He aimed to bring mysticism into the field of science as "open-minded, rigorously tested, rationally coherent, and illuminating."[86a,vii] His approach was empirical: mysticism was taken to be "the study of everything non-physical, including the other worlds and their archetypal governance, as well as our spiritual bodies, the facts and their relationship being known by the self-evidence of direct observation and not by reasoning or speculation."[61,1]

There are several points to note in this definition. It views mysticism as a fact-finding study, and so should be treated as a branch of science, based on direct observation and experience. Speculation and theory-spinning is put out of court; this is radical empiricism.

In the definition there is no mention of formal religious practice or a particular deity. Whiteman was critical of religious beliefs and terms that are not directly warranted by experience. Nevertheless the phrase "archetypal governance" points to experience of what he termed "Ultimate Wisdom" or "Source of Right and Good",[86a,20] transcending usual physically-based experience. A scientific mysticism will have to throw light on this experience.

Another point about the definition is that it explicitly refers to the non-physical, to "other worlds" of conscious existence, not the physical world generally thought to be the only "reality". To conventional thinking, anything beyond the physical tends to be inconceivable, not open to scientific study. It must be the task of scientific mysticism to make the idea of "beyond the physical" not only conceivable but unavoidable in the face of evidence.

Lastly, if mysticism is the study of "everything non-physical", then it could be said to include the subject matter of psychical research, which also deals with phenomena not open to physical explanation. Marshall (2011, 2015) has recently emphasised the close relationship between the psychical and the mystical. Demarcation between the two is nevertheless needed. Whiteman saw mysticism to go further than a study of telepathy, clairvoyance and other psychic (or psi) phenomena; mysticism was seen to involve a "sense of ultimates", "openness to guidance", and "transformed being"[86a,42] as underlying principles, discussed in Chapter Six.

He recognised that finding these principles can be done effectively only by taking adequate account of general science, from the quantum level to the affairs of ordinary living. At the same time his personal experience and wide knowledge of psi phenomena led him to see these phenomena as integral to mystical experience, as well as being an inseparable part of science. His syncretic view was of a broad science that included psi and mysticism without being self-contradictory, and the aim of the present book is to outline this *scientia.*

It could be said that one of Whiteman's failings was his use of the word "mysticism", with all its confused meanings. Yet he saw his usage to be in accordance with long-established Eastern and Western tradition, grounded in "the self-evidence of direct observation," as stated in his definition. His own grounding spanned mathematical physics across to psychology, psychical research, the wisdom traditions, and was informed by his own psychic and mystical experience. His notebooks describe over 7000 of his own experiences. The breadth and style of coverage was revolutionary.

His first book, *The Mystical* Life[61], has the subtitle, *An outline of its nature and teachings from the evidence of direct experience.* The word "evidence" also appears in the title of his three-volume series, *Old and New Evidence on the Meaning of Life.*[86a,,00,06] His methods rested on observation and conceptual analysis of what he termed "the inner constitution of nature". This is a phrase from the title of his book,

Philosophy of Space and Time and the Inner Constitution of Nature,[67] which was subtitled *A Phenomenological Study*..

Phenomenology was understood in Edmund Husserl's (1931) sense to be the rare gaining of "face-to-face self-evidence and detailed precision", shedding the usual "cloak of ideas" that covers perceptiveness. Science in this context is not the conventional exercise of trying out theories (guesses) about how things work, but in discerning the logic in experience and in the inner constitution of nature. Scientific mysticism tries to apply consistent face-to-face phenomenological analysis, as discussed in Chapter Eight. Evidence is gathered from claimed observation and direct experience of states perceived to be not ordinarily physical or worldly. The evidence may be taken together with evidence from psi research, also not dealing with "normal physical" happenings. This requires an open-minded examination of experiences that have a sense of phenomenological reality, yet cannot be referred to what the bodily senses detect, either directly or indirectly, and cannot be attributed to aberration or pathology. Whiteman used the term "non-physical" for such experiences, and for states and observed objects that are "not locatable in physical space".[94,84] A scientific task, then, is to clarify what is meant by "not locatable in physical space", and to discover principles underlying non-physical experience.

A prime underlying principle is Whiteman's many-layered conception of space and of time. In correspondence with Ian Stevenson, a psychiatrist deeply involved in the study of reincarnation, Stevenson wrote of becoming "more and more convinced that a further understanding of the existence of two spaces, or perhaps multiple spaces, is necessary for our understanding of the relationship between minds and brains and also for the solution of many problems in parapsychology." (10/12/1986). Whiteman's reply (12/1/1987) was that "everything hinges for me on the admission of 'other spaces'." Given the difficulty most people have with the idea of experiencing a non-physical space, a study of mysticism and psi has to dig into what can be understood by "other spaces" and what led Whiteman to his understanding. This involves dismantling what he termed "one-level naturalism", the idea that all things, events and their causation happen only at the level of a physical world that we sense with our bodies. Without correcting this notion, an effective scientific mysticism and a theory of psi was held not to be possible.

Essential to the mystical and psychical tasks are skills of freeing the individual's mind from normal physical life and entering states of awareness different from the physical, advocated by Husserl and discussed from Chapter Four onwards. This line of thought is foreign to standard Western ideas, and needs close examination if it is to be accepted into scientific study. At a deeper level this reaches the matter of religion. To prepare himself for this investigation, Whiteman became fluent in classical Greek, Biblical Hebrew, Vedic, Sanskrit and Pali, the latter being the language of Buddhist canonical writings. He made a close study of thinking from Minoan, Vedic and early Hebrew times up to the initial period of early Christianity, after which he saw religious thinking corrupted by dogma and theorising in both Western and Indian texts. In his analysis, "nothing is presupposed," and theology simply meant "the application of the phenomenological method to our awareness of the Divine."[06,vii] So, "the word 'God' *(Theos)* must be taken to stand for the Archetypal Reason in all."[06,64] This empirical approach to religious thinking is a basic requirement of scientific mysticism.

The idea of "the Archetypal Reason in all" is a cornerstone in Whiteman's scientific treatment of mysticism. Quantum theory provides an essential element. It is held that an individual's observations are underlain by causal substructures that are not physically located, discussed in Chapter Two. An individual's own potential for observation combines with more universal potentialities to deliver an actualisation on an occasion of observation. What the individual observes will depend on his/her state and abilities. Put the other way, the state and ability of an individual will determine what is observed. What is manifested to an individual in a dissociated or non-physical state will differ from what is seen by someone in a "normal physical" state.

If the causes of what we observe physically are not locatable in physical space, then the observed world is not the primary "reality" we unthinkingly take it to be. It is a product of underlying realities, which are not physical. This is a core understanding in mysticism, and indeed in what is now known as post-materialist science (Beauregard *et al.* 2014). In this mode of science there is a need to examine the causal background of what is delivered as physical and as non-physical experience. A hierarchical view of causality reaches back nearly three millennia to the Upanishads, and is a basic ingredient of Whiteman's scientific mysticism. Timeless elements of causation, of *essence,* were contrasted with elements of *existence,* the actualisation in time of timeless potentiality, discussed from Chapter Eight onwards. Any event or idea needs to have these causal constituents recognised before it can become intelligible at the level of discerning "the archetypal reason in all". The relationship between potentiality/essence and actualisation/existence is central to Whiteman's analysis of "the inner constitution of nature."

He saw the need for a scientific mysticism to penetrate even further along lines started in the Upanishads. There, a sixteen-fold cycle of conscious functions was recognised, the *kalās,* "operational parts in creation".[86a,201] He recognised that these "subjective" functions match a sixteen-fold "objective" cycle of elementary particle physics, quantum field theory, and a sixteen-fold universal system based on three space and three time dimensions. This matching gave his scientific mysticism the power to integrate physics with psychology, studied from Chapter Nine onwards.

Whiteman's analytic psychology had roots in early Buddhism and the Yoga Sutras, particularly with the concept of what he developed as a "corporate structure of personality". Through his own experience and various Eastern and Western sources he recognised an enduring core to an individual, which was accompanied during physical life by a fluctuating population of "contributory minds". A task of the individual was to recognise the core which transcends physical life, and to deflect influences especially of undeveloped co-minds crowding around in the physical life. Failure to do this can lead to catastrophic events which Whiteman closely examined (Chapter Twelve onwards). Understanding spiritual development was seen by him to hinge on understanding the causal structure behind the individual; in fact without taking into account a corporate structure of personality, an effective science of mysticism and psi seems hardly possible, as later chapters will repeatedly show.

Taking all this into account, from the individual to the universal, what may be expected from an integration of mysticism and psi, bedded down in science, is a journey through some of the deepest layers of thought. It calls for an open mind and willingness to engage with ideas that may at first seem inconceivable to westernised conventional thinking. This style of thinking should be accounted for in a scientific treatment of mysticism and psi. Conventional westernised thinking along materialist lines may be seen to be aberrant in the context of world history. A product of a peculiarly Western development, it became entrenched during the period of the Enlightenment in the seventeenth and eighteenth centuries, founded on the one-level naturalism noted earlier. As discussed in Chapter Eleven, it had a fragmenting effect. Following Whiteman's revolutionary synthesis, the success of this book may be measured by the depth of insight it is able to bring to the relationship between science, mysticism, and psychical research as complementary rather than contradictory. It is an insight that annuls a one-level mode of thinking.

CHAPTER TWO

POTENTIALITY AND ACTUALISATION

It seldom occurs to us that every viewing of the world is a remarkable, unique event. In a room we see a chair in front of a table. Move round the room and we see the table in front of the chair, a completely different scene. What makes us think the chair and table are "the same" from either view? We realise that we are not faced just with a random assemblage of different appearances; the different perspectives seem intelligible, and what makes them so is that behind them all must be a continuing logical and creative structure which is the origin of all appearances from all viewpoints. But, Whiteman argued, this logical structure cannot be contained in physical space in the way a single appearance seems to be. The behind-the-scene substructure must continue to exist while we view different perspectives, otherwise we should have no ground for calling the furniture "the same". Yet it must belong to a sphere of potentialities that are not a part of the physical universe of appearances, of actualisation.

The idea of substructure, of potentiality and actualisation, is a key to understanding Whiteman's thinking. The idea had been worked over by the physicist Werner Heisenberg in his *Physics and Philosophy* (1959). He saw that a physical "happening" is an occasion of observation, open to physical description. But the possibilities or tendencies governing the happening are not observable in the same way; they are not physically locatable, at best handled as mathematical constructs. So lying behind a physical event, behind any viewing of the world, one has to recognise what Heisenberg (p. 56) called "an objective tendency or possibility, a 'potentia' in the sense of Aristotelian philosophy." And the governing potentia, the "reason" for something to happen, is abstract in a physical sense: you can't put reason in a test tube or find it sitting on one of the chairs.

For Whiteman the term "reality" could apply only to this underlying reason, potentiality, not to appearances, to actualisations. Appearances are manifested only on an occasion of observation. The demand of quantum theory, as he saw it, is that "the 'purely physical' is incomplete, and manifestation requires that the circumstances and mental powers of an individual observer can be given before anything definite can be manifested."[00,132]

The idea of an observer-dependent world might seem ridiculous, as it did to Samuel Johnson when he kicked a large stone, saying of the idea, "I refute it *thus.*" (Boswell 1791, p. 471) He was in fact actualising the stone in a very drastic way. Does the stone exist when no-one is seeing or kicking it? The *potential* for it to appear physically does, yes; but does the physical actuality? Literally, no; there are only general and highly complex non-physical potentialities,[75a,195] which will be actualised differently for different people seeing or kicking the stone from different angles. The potentialities will exist whether or not they happen to be actualised by any individual, provided they remain undisturbed, yet we are apt to think of actualised, tangible objects as themselves being self-existent and placed permanently external to us. Whiteman saw this as an error that was pointed out long ago in the ancient Indian doctrine of *māyā,* the "trick" of creation[67,148].

Māyā has two powers or functions, he noted. By the power of projection, potential relations become presented as things, actualised in the observer's world. The observer is in a sense a creator of what he sees, and what he perceives can be said to be a product of his observational status. A drunk will see different things from one who is sober; even more radically, an observer who is freed from the mode of physical perception will actualise for himself things that are beyond the reach of ordinary physical vision. This is the beginning of psychical and mystical perception.

The second power of *māyā* is enclosure. The perceiving self becomes closed off, giving things a false, out-there, non-mental quality, which leads to a supposed separation of mind from non-mind and to the problem of how mind can somehow connect with supposedly non-mental things. Yet as Heisenberg pointed out (p. 75), the interplay between nature and ourselves "makes the sharp separation between the world and the I impossible." The idea in quantum theory of the indivisibility of the phenomenon of viewer and viewed is evident here; as Whiteman put it, "nothing specific can be actualised without the integration of 'subjective' and 'objective' factors, not in succession, but on the instant of actualisation."[86a,240] There is joint actualisation of the "impersonal" potentialities of the event and the "personal" potentialities of the observer. The resulting interplay between physics and psychology is a major theme in Whiteman's work, to be discussed in Chapter Nine.

The power of enclosure gives rise to the philosophical tangle of dualism, of mind and matter, but here we are concerned with the duality of potentiality and actualisation. It holds that when an observation is made, all potentialities—from the universal to the individual—are integrated on that occasion to produce a unique actualisation, attended by the powers of *māyā*. The state of the observer is itself a part of the potentiality governing any appearances, so the process of actualisation will be different for different people. Someone conscious in the physical world will produce a physical actualisation. A person during an out-of-body experience, near-death experience, or psychic or mystical experience will produce an actualisation in the particular world his mind is functioning in, which is not the physical world actualised by what we call "normal" sensing. So, as Whiteman wrote, "The difference between one 'world' and another then depends upon the state of life in which the observer happens to be settled."[06,3] This is an experiential basis for the idea of "other worlds" of conscious existence, which is a fundamental part of Whiteman's definition of mysticism (Chapter One).

This leads back to the question, what is "reality"? The materialist idea that physical matter is all that is "real" can be seen to misunderstand the powers of *māyā.* The potentialities that deliver a stone on being kicked are not part of the stone's physical world, yet they must be "real" to bring about a stone that stubs your toe. There would be no actualisation of a physical stone if relevant non-physical potentialities were not real, including mental potentialities that bring about the stone-kicking. It is an idea that takes some getting used to. It means departing from the conventional error which, as Whiteman emphasised, "arises from inability to see that externality, however 'real' it may seem to us, comes into evidence only with actualisation ... and previously *did not exist as such.*"[67,148] An individual's percepts are actualisations of such potentiality-fields, actualisations which result from integration of objective potentialities with his own individual potentialities, including subconscious elements[73,355].

This is not to downgrade physical actuality. One could say that lifting up the stone and moving it physically will alter the potentiality to stub your toe. In that sense, actualisations as well as potentialities could in their own particular ways be called real, even if actualities are ultimately conditional. The relationship may be seen as a reciprocal, integrated feed-back one; Whiteman insisted on the importance of clear-headed, experience-based evaluation of physical presentations, necessary for understanding potentiality-actualisation relationships as part of a greater whole of comprehension. And further, to recognise that realms of potentiality have the capacity to deliver different levels or worlds of physical and non-physical actuality, depending on the state of the observer. Comprehending different levels or spaces is a starting-point for the scientific study of psychical phenomena, as Stevenson noted; it is also a starting point for a science of mysticism.

CHAPTER THREE

PHYSICAL AND NON-PHYSICAL

In his *The Mystical Life*[61], as noted in Chapter One, Whiteman wrote of mysticism as "the study of everything non-physical, including the other worlds and their archetypal governance ..."[61,1] One needs to know what he meant by "non-physical", and this means carrying out one of the first tasks in science, which is to classify. This is what this chapter mainly deals with.

A classification in itself is not right or wrong, only more or less useful for the purpose at hand. A useful classification arranges material in a way that allows comprehension and the ability to focus on particular features. Probably the most basic distinction that needs to be made in the scientific treatment of mysticism and psychic phenomena is between what could be classified as "physical" and what is "non-physical". It is fashionable nowadays to replace the term "non-physical" with "non-local", taking a cue from the idea of non-local entanglement in physics. The term "non-local" is not appropriate in a study such as this, where Whiteman wrote of "located non-physical objects" in non-physical space.[86a,7] Replacement of "non-physical" with the term "non-local" leads to the contorted phrase "located non-local objects." This is discussed further in Chapter Nine.

Whiteman's handling of the distinction between physical and non-physical changed slightly over time; the latest meaning he gave to the term "physical" pertains to "normal awareness, existence, or state of conscious life, as seemingly located in our familiar world and clock-time."[06,4] The phrase "seemingly located" refers to the powers of *māyā,* discussed in the last chapter. The term non-physical then means "not having some quality or qualities essential to being 'physical'." Evidently he assumed that the idea of "normal" and "familiar" were reasonably operational. He meant by "normal" the state of "absorption in physical phenomena, and absorption in discursive thinking ... merged in a general impression of the physical world and body."[73,355] There is a tendency for these definitions to go round in circles: "physical" means "physical awareness" which means "physical". But, as will be seen, the feature of clock-time seems diagnostic. Release from clock-time requires detachment, some shift from physically-based consciousness and perception, as discussed in Chapter Nine.

Whiteman noted almost in passing that "gravitational effects are quite different in the spiritual worlds."[06,xxix] This could be explored as a way of distinguishing "physical" from "non-physical", together with time effects.

Uniquely difficult to resolve is the physical/non-physical classification of elementary particles, fields, wave-functions and quanta. They are not located in our familiar world and clock-time, Whiteman argued, because "none of these can be observed directly," so they are not classifiable as physical. He wrote, "Whatever view we form of them is conjectural, and imaginative in character. The imagination that we have concerning them could be described as non-physical. But as conceptually created structures they must count as auxiliary mental creations in a physical state, and not as objects of attention in a stabilised non-physical state."[00,12]

Practicing physicists generally take these items such as particles to be "physical", but as Heisenberg noted (1959 p. 59), they may be viewed "rather as a kind of 'potentia'." To the extent that the terms relate to potentiality, they connect with what is real, yet being "conceptually created structures" they are not actualities at some point in physical space-time. "What happens is that a particle-like effect is manifested from the world-filling potentialities only when and according to the way in which an observing-apparatus is set up." (Letter to Scott Rogo, 15/6/1978, cf [00,12]) As such they may be referred to a "thought-image sphere" discussed below, which is mental, not physically located.

Whiteman pointed out that, perhaps unexpectedly, the usual conception of "reality" being restricted to a physical universe is undermined by physics itself. Quantum theory shows that waves or other structures controlling probability are beyond the scope of physical observation, are not located in physical space.[73;75c] Yet he did not suggest that quantum theory itself could help explain psychical phenomena. He wrote that these phenomena "are reasonably held to be not 'physical' and thus not within the province of quantum theory." The special "epistemological relevance" of quantum theory to psi phenomena had to do with its adoption of a multi-level "hierarchical world-view",[73,357] notably the levels of potentiality and actualisation.

Developments in physics continue to show conditions where explanation cannot be made in terms of physically located matter, but in "non-locality" as in the case of quantum entanglement. The picture here is that two new "particles" produced by the decay of an original particle have different quantum spin values; if the value of one is determined, the value of the other correspondingly changes independently of physical space and time. The cause of what is observed physically cannot be observed physically.

Happenings such as sighting apparitions raise other difficult physical/non-physical problems of classification. There is enough evidence to show that apparitions can be "real" to a viewer in some sense (e.g. Tyrrell, 1953, Kelly *et al.*, 2007). The influential philosopher Immanuel Kant rejected their reality on conceptual grounds. His rather tortuous argument boils down to the objection that apparitions could not displace physical matter, therefore they are not physical; and if they are not physical then they cannot be "real". This was an expression of the prevailing one-level naturalism, discussed in Chapter One, which still dominates thinking. His views will be discussed in Chapter Eleven.

Whiteman's view was in agreement to the point that an apparitional sighting could not be said to be in physical space; a physical actualisation was not conceivable. Initially he classified the appearance of an apparition as a non-physical "opening" into a normal physical space, physical and non-physical regions being observed at the same time.[00,105] But his views changed. In a letter to me of 11/5/2002 regarding his 2003 paper[03a] he wrote, "I certainly rule out absolutely the idea that part of a spatial field can be normally physical and other parts definitely not physical." In "openings", discussed below, the apparition as well as the surround are "*not* normally physical, the mental state being dissociated." Further, "I would now say that in Opening there is a division of the spatial field into a central focal part which is definitely non-physical, the scene being different from what would be normally seen, while the part outside the focal area is duplicate physical." The term "duplicate" carries complexities discussed in Chapter Six onwards.

He cited from Tyrrell's *Apparitions* (1953) a case where two young men described an apparition entering their bedroom. He rejected the view that the sight of a figure entering through an open door (when the physical door was closed) was simply imposed on the physical scene. There was no report of "where a dividing barrier could be placed between the light-and-space world that was physical and a portion of it that was non-physical."[03a,101] In these and other sightings he concluded that the percipient "is actually dissociated to a small degree. Intelligible factors behind the physical scene have combined with factors of perception and creativity in any available beings (whether physically living or not) to produce a *conjoint* manifestation upon occasion, dissociative in character throughout."[03a,101] The state of dissociation which may induce an apparitional sighting is discussed below.

All sorts of variations may arise, such as the apparition of a living person being viewed without the knowledge of that person.[86a,112] The apparition in such a case was considered to be a co-mind or double, discussed in Chapter Fifteen. But none of these phenomena can begin to make sense unless the distinction is recognised between what is physical and what is non-physical.

CHAPTER FOUR

CLASSIFYING NON-PHYSICAL STATES

If the task of distinguishing "physical" from "non-physical" is complex, even more so is the task of identifying and classifying non-physical states in their "inexhaustible variety".[86a,8] A scientific classification of non-physical states, Whiteman pointed out, requires there to be a stabilised condition to allow investigation, lasting at least a few seconds. He recognised "five chief types of stabilised non-physical state",[86,9] five fundamentally different ways in which one can pass from a physical state to a stabilised non-physical one.[86a,8] Being a classification, incipient, transitional or hybrid types have to be allowed for, but the five general states are worth looking at closely.

Two of these non-physical states, opening and dissociation, refer to experience in which there is no impression of being located apart from the physical body, or having a bodily organism separated from the physical body; the remaining three involve awareness of separation from the physical body, commonly referred to as "the out-of-body experience". The word "the" implies only one general state, but Whiteman's classification corrects this simplification by describing three main different states. He avoided the term "out-of-body", since the experience is usually an embodied one, although not in a physical body. He used the neutral term "separative experience" instead of "out-of-body experience".

Of the two non-separated states of opening and dissociation, opening is perhaps too broad a category. It includes visions and other sensory effects, or states when consciousness is focused in a mental space, as in imagination, musical or other modes of creation. It also includes apparitional sightings and other psychic or (more intelligibly) mystical experience. There is no impression of being "out" of the physical body[68,9]. In cases such as visions or viewed apparitions, what is perceived cannot, as discussed above, occupy strictly physical space; the opening presents location in a non-physical space perceived psychically by the physically-located viewer.[68.92]

He noted that sometimes "one can pass through the Opening, as it expands, to a separative state."[03a,101] This was considered to have occurred in reports of near-death experiences.

Whiteman's interest in the state of dissociation was as a prelude to non-physical states, rather than as some form of mental disintegration. In a state of dissociation, the whole field of awareness may have a non-physical character, as in deep thought, but there is "no clear impression of being in a stabilised bodily form apart from or different from the physical one."[68,10] In a normal physical state there may be "very brief episodes of dissociation, when the physical world is as if blotted out in thought or clouded over in imagination, and there is loss of all sense of bodily form."[03a,102] In light dissociation the state might be misconceived to be normal physical, yet, as in the case of post-hypnotic suggestion, the subject's perception does not match what is physically perceivable.[00,144] Light dissociation, without awareness of the state being altered, was held by Whiteman to occur also when viewing an apparition.[00,429]

In longer term he regarded dissociation as "transitional in character, between a normal physical state and separative states, on the one hand, and states of habitual irrational addiction on the other hand."[03a,101] "Lower states" of dissociation included disorganised states of emotion, neurosis and psychosis, and may be involved in psychological breakdown. "Higher states" involved "holistic non-attachment and higher rational direction (to the right and good, or essential morality) 'in the world but not of it', leading to a broad sense of unitive transformation and fulfilment."[03a,101] Also in dissociation there is participation in spiritual "skills" (Chapter Nine) which free the mind from normal physical life, entering inspirational or mystical states of consciousness different from the physical state.[00,84] He examined "lower" dissociation in the murders discussed in Chapter Twelve and after.

In addition to opening and dissociation, three non-physical states were grouped as separative or out-of-body experience. In contrast to opening and dissociation, all three were characterised for him by "a sharp and unmistakable distinction between experience *as from the physical body* and experience when having fully *entered* a non-physical sphere of existence."[75b,100] Studying separative experience turns out to be extremely complex, as the next three chapters start to show. Whiteman noted that the usual conception of an out-of-body experience is something akin to a surgeon removing a kidney and momentarily placing it beside the body. The separated entity is in the same space as the physical body. Yet an "out-of-body body" is not detectable by physical means in the way an isolated kidney is, therefore according to Whiteman's criterion it cannot be regarded as physical, and its location can only be in non-physical space. This is a crucial point, and the idea of experiencing different spaces raises a host of conceptual problems, considered in the following chapters, notably Chapter Seven.

The greatest of these problems is perhaps the challenge to conventional one-level naturalism, the idea noted in Chapter One that there can be one space only, physical space, in which all things, events and their causation happen. In addition to separative experience, other phenomena studied in psychical research present the same challenge to one-level naturalism, according to the way Whiteman interpreted them, because there is involvement of some non-physical state, be it only through light dissociation. These are usually classified as telepathy and unconscious presentiment, clairvoyance or remote viewing, psychokinesis, and precognition. To account for these, Whiteman drew on the idea of potentiality and actualisation as it relates to individual awareness.

The potentiality-actualisation principle was seen to operate in the mentality of every individual. Several hierarchical levels of potentiality and actualisation were thought to be involved. Whiteman considered thought and images to be actualisations in the consciousness of an individual's "thought-image sphere", as he called it,[73,355] which includes memory and imagination. From the point of view of physical actualisation, a thought-image sphere is part of a potentiality field involved in bringing about the physical happenings; but from the point of view of an individual's conscious awareness, the sphere is a field of actualisation. Its potentiality fields include subconscious elements and physical potentiality.

So telepathy was seen to result from the interaction or resonance between the thought-image spheres of different individuals; these spheres are not physically located and so allow transfer of information independently of physical space and time, bypassing physical sensory structures. Clairvoyance is resonance between the potentiality-sphere of a physical thing or event and the thought-image sphere of an individual, resulting in the production of impressions in the individual's thought-image sphere. Once again, this happens independently of physical space and time, and so bypasses physical sensory structures. Psychokinesis may be regarded as a reverse clairvoyant interaction; potentialities, linked mostly in some way to an individual, result in physical actualisations without involving direct physical mechanisms.[73,357;75c,130]

Potentiality-actualisation relationships are not to be seen as static; potentiality-fields continually change or develop, have feedback loops, although not in physical time. This will affect the outcome of future physical events, which helps to explain how precognition happens.

Whiteman saw precognition as actualisation in an individual's thought-image sphere of potentiality-fields that may affect future physical events. He/she becomes directly aware of the potential for some event to happen in the physical future. The individual may even intervene in the physical actualisation of the event by altering potentiality-fields, like moving an infant away from a place where an accident is "seen" to be threatening. This allows prevention of a precognised event from happening. It also accommodates the idea of free will. As discussed in later chapters, even if potentiality-fields are time-ordered, resulting in a succession of events actualised in the physical world, the ordering is not fixed in the way a road map is fixed. It is accessible in states of physical dissociation, and so allows limited alteration and the exercise of free will.

Following Whiteman's thinking, these phenomena can begin to make sense only if a one-level naturalism is jettisoned. As he wrote to Stevenson, "everything hinges for me on the admission of 'other spaces'." This has wide implications. As Carr (2015, p. 260) argued, these phenomena "all require some form of space which is *communal* in the sense that it can be accessed by other minds." If mental phenomena require a communal space, it is, Carr (p. 265) noted, "tantamount to positing some form of Universal Mind..." This will be a constant theme in this book.

From the point of view of mystical development, Whiteman regarded separative experience as by far the most important among all these experiences, and certainly the most complex. This will be explored in the next two chapters.

CHAPTER FIVE

SEPARATIVE EXPERIENCE

Chapter One reported Whiteman's remark to Stevenson that "everything hinges for me on the admission of 'other spaces'." This ties in with his definition of mysticism (Chapter One) as "the study of everything non-physical, including the other worlds ..."[61,1] In Chapter Two it was stressed that, for a full understanding of potentiality and actualisation, we need to recognise that fields of potentiality have the capacity to deliver different spaces or worlds of physical and also non-physical actuality. What is actualised is observer-dependent. Without this understanding, psychical and mystical experience cannot begin to make sense.

A direct experience of other worlds, Whiteman maintained, occurs most effectively in separative or out-of-body experience, when non-physical spaces or worlds are entered in a separated state. The psychological and neurological literature on purported out-of-body experience generally takes little or no account of the kinds of experience treated by Whiteman, or of the causal principles behind the experience; at best the experience is regarded as hallucinatory or pathological in conventional terms. A notable exception is a paper by Neppe (2011) that points out the common mistake of treating reports of out-of-body experiences under one single etiological umbrella. He suggested a "multi-etiological model" in which causes and explanations of recorded experiences are classified into four groups: psychological aberration, neurological dysfunction, psychopathology, and paranormal experience. He noted that paranormal experience appears phenomenally quite different from cases studied under the first three groups. Whiteman's treatment focused on these experiential aspects; he spent almost no time on the other categories identified by Neppe.

A great deal of study has been devoted to out-of-body experience, astral travel or similar terms (e.g. Kelly *et al.*, 2007; Nicholls, 2012). Whiteman's term "separative experience" (Chapter Four) has not been taken up in the literature, but it is used in this book as a more accurate and logical term. The commonest type of experience was classified by him as "primary separation". There is no awareness in the physical body;

"another world of non-physical space and time is presented, as 'real' in its own way as the physical world, autonomous, and not at all subsidiary to the physical world as are thought and imagery."[86a,57]

In "secondary separation" the physical body is under some degree of conscious control, but principal awareness is in another organism seemingly not located in the position of the physical body, and the two bodies are quite independent. The experience is doubly-conscious; the usual term "out-of-body experience" is particularly inappropriate, since there is awareness "in" the physical body as well as the separated body. The same is true of "tertiary separation", where the physical body is normally under complete control, yet there is full awareness of being in a different bodily form, positioned as if in the location of the physical body. Differences between the two bodies may be radical, such as one being male and the other female. The ontological significance of this will be discussed later.

Experiences recorded by him may clarify primary, secondary and tertiary separation. First is a primary separation. These tend to be relatively prosaic.

> In the early morning, while lying in bed half-asleep, I suddenly became keenly aware of an inner call to go to the front of the house. The state then was clearly one of dissociation, with a lack of clear perception, but quite clear of fantasy. I got out of bed non-physically in what seemed to be my normal bodily form, and walked along the passage (in a non-physical "duplicate" of the house) to the dining room. On passing through the archway which led to it, I turned left, looking towards the sideboard, and all at once the awareness of being in separation became much more clear and precise.
>
> The figure of Nancy B. was standing at the sideboard, back to me, apparently looking for something on top [she used to come to the house to teach music, and sometimes there were letters or messages left for her there]. She turned, and I could see that her appearance was as in physical life (she had died a few days before), though her face seemed pale and drawn through recent suffering, quite different from its usual relaxed expression.
>
> I moved up to her, saying, "Is it you, Nancy?", and put my hands forward so as to hold both her arms. I had no doubt that it was really Nancy, and said, "This is wonderful! I admired you so much." There was an impression as if she was rather lost and not fully comprehending the situation. I leant forward and kissed her, and as I did, there came a tremendous sense of personal contact and reality. She then touched me similarly on the arms, and I sensed a question arising in her. I replied, "Yes, I am quite solid!" After a short period of reflection she seemed to say, "You too?" This surprised me, but I understood it as expressing some

realisation that the state was beyond death, and puzzlement over the fact that I seemed to have passed over too. My mind then seemed to come upon a rather confused state, though I was able to observe the confusion without at first losing control ... A dreamlike imagination seemed to take hold of me, of difficulties which might arise if one came back alive when one had been supposed dead.

The words sprang up, "What will you do when you have to get a new passport and things like that?" ... I was still able to realise the absurdity of such questions, but the growing state of confusion led presently to resumption of the physical state. There was an almost imperceptible fusion of the two states, so that something of the experience remained vividly with me.[03a,103;06,33]

An abbreviated example of a secondary separation:

About 2 p.m. I was lying down, face upwards, and reflecting on certain philosophical aspects of the General Theory of Relativity, which I had been lecturing on in the morning. Suddenly it was as if a curtain was lifted in front, and there appeared a small group of persons seated at a table as if in committee... The scale was normal and I appeared to myself to be standing alongside the table on the left side and thus only a few feet away from the people... After taking in these facts, my eyes were drawn to the person sitting on [the chairman's] left and in whose direction I was facing. It was a woman, possibly about 40 years of age, whose gaze was directed to me with extraordinary intensity, kindly though serious and steadily intent...[86a,101]

He later discovered that the figure resembled a portrait of an early psychical researcher, Eleanor Sidgwick, died in 1936, who, like him, was a Cambridge wrangler (first class in the mathematics tripos). This cannot be read simply as a dream since he experienced the "difficult to describe" nature of the conversation, which "in such states is a matter of direct communication of thought and feeling, and the thoughts, moreover, are put into the words of the one who receives them."[86a,101]

An example of tertiary separation is in a transformed state, characteristic of that state:

Revelation of the acceptable loving form for more than ten minutes, while lying in bed, fully awake. Vivid appreciation and admiration of the living contours and proportions, with their inexplicable aesthetic perfection; not by self-will, but by loving surrender. No piecemeal modifications of the existing lower form, to which the beauty within bears no resemblance, but the manifestation of an acceptable soul-life replacing the inacceptable. Joy and gratitude at *being* that form and responsive life fills the heart. No sudden end; but the sense of remaining in that soul-form and with that love remains.[61,221]

In states classified by Whiteman as tertiary separation, the non-physical body holds much of the focus of attention but does not appear to be spatially apart from the physical body; "the locations of the two bodies seem the same, in spite of differences in their form and/or size and possibly also in their surroundings."[86a,13] He did not leave a description of how this may happen, but it seems to tally with a view widespread in spiritist circles that different "levels" of experience and existence all coincide spatially but have different "vibration frequencies", so there is no contact or communication between the various levels except through special mediating conditions. All four of the other non-physical states described by Whiteman could be included in this multi-frequency system, which could also be interpreted as an interpenetrating multi-space system.

This encroaches on theory-spinning, an area that Whiteman carefully avoided. Nevertheless, he did include "interpenetrating spaces" in a particular set of "facts of experience". These facts "do not fit into the ordinary frame of physical experience, and yet must be admitted on the ground of rationally coherent testimony, continually repeated, from many and varied sources whose sincerity and clear-sightedness are not usually regarded as open to doubt."[61,224]

A recorded case of his own was when, standing quietly in his bedroom, he entered tertiary separation: "I knew myself then in my spiritual body of love, standing similarly and without apparently having shifted my location, but in a boundless world of intelligible space and substance."[06,310]

He did not present the idea of interpenetrating spaces as some kind of cosmological system; it had to do with the ability of our stream of consciousness "to occupy a perceiving organism in several spaces simultaneously—rather as we can think and see in the physical world simultaneously, and likewise reflect on our thoughts, feelings and perceptions, both interior and exterior."[61,24]

He noted that all five types of stabilised non-physical experience—opening, dissociation, primary, secondary and tertiary separation (Chapters Three and Four)—may range in reality-impression and intensity from being weak to seeming even more real and meaningful than in physical life. The commonest report of primary separation is in fact often accompanied by a statement that the experience felt "more real" than physical experience, and meaningful to the extent of being more or less life-transforming. At its most powerful, "Everything in such states, spatial characteristics included, is known as if in archetypal unchanging form, and is therefore startlingly more real (rationally-objective, ever-present) than the derivative and shifting forms of the physical world."[86a,143] On the other hand, in undeveloped cases the mind might not be fully reflectively conscious, so "the phenomena, though reflected on, are unpleasant or seemingly irrational, and not under such control as can give us peace of mind. I call the condition, then, *fantasy-separation*."[61,17]

But however compelling the experience, there remains the question in the mind of the outsider: how "real" are these experiences? In what way could they be "real", being so different from normal physical life? And what actually is meant by "real"? These are fundamental questions that a science of mysticism has to tackle.

CHAPTER SIX

REALITY OF SEPARATIVE EXPERIENCE

This is a topic of immense complexity, but I have suggested (Poynton, 2001) that in out-of-body research, one may take "real" simply to mean what passes the test of being scientifically usable. One needs to ask, then, whether any records contain scientifically usable data that allow one to analyse and judge whether the experience was real in the way claimed, rather than a result of inaccurate observation, dishonesty, imagination or hallucination.

A usual criterion for the "reality" of the experience is the presence of physical corroboration: the experiencer "saw" something in the physical world that he or someone else can confirm was there. In the academic courts the jury is still out regarding this. The observations could have been made by clairvoyance or remote viewing, and do not necessarily mean that something somehow separated from the physical body to make the observation. Whiteman made this point as well, although, as discussed in the next chapter, the issue is far from simple.

The source of evidence therefore has to come entirely from assertions about an experience, which raises questions about the "scientific admissibility of verbal description."[03b,41;06,6] Whiteman considered this under the heads of authenticity and reliability.[86a,35;06,7] Authenticity involved the manner of recording the experience (detail, length of time elapsed since the experience), and mental capacity of the experiencer to make a report that is scientifically usable. Reliability concerned the ability of the reporter to check details of the experience, and the degree to which familiarity with separative experience was shown, strengthened by similarity with reports from other experiencers. Whiteman's wide range of his own experiences, when added to his collection of similar reports from many other sources, led to a high personal scoring on reliability. His detailed and immediate diarising of his experiences gave a similarly high scoring on authenticity.

This left him in no doubt about the reality of separative experience, so he set about developing his own criteria "to distinguish between what is deceptive or illusory and what has the quality of face-to-face objectivity and truth which we describe by saying that a thing is *real*."[86a,33] Yet he saw complications in this task, and his criteria for the reality of separative experience are not what occur at first to most people.

Most people settle with a phenomenon being either "real" or being "not real", but in this realm of experience he saw a need to evaluate the *degree* of reality involved. This is because non-physical experience may range in reality-impression from what seems imaginative to what can be judged even more real and meaningful than normal physical experience. This requires setting up a numerical scale, which he called a General Index of Reality,[86a,42;06,10] reproduced on page 26.

Cardinal requirements of the General Index are continuity of memory between physical and non-physical experience (M), free observation (R), ability to compare physical and non-physical states accurately (H), communication of thought with others (P), and a feeling of transcendence of physical life (I) and awareness of higher significance (V). The table includes four functions of creativity, symbolised by four directions of space, discussed in Chapter Ten.

The index covers a range from uncontrolled dreaming (zero rating) to what is experienced as spiritually releasing with a high degree of reality, namely mystical experience (high rating), which score on items I and V. The inclusion of dreaming with mystical experience in this scale might seem to bolster the idea that the whole scale lies in the realm of vivid subjective fantasy. But even if dreaming is conventionally supposed to be a kind of private event manufactured somehow in the brain, the experience itself cannot be referred to the physical space of the dreamer's physical body; it is happening in a different world, physical-like, but not physically located. To this extent it is an out-of-(physical)-body experience, but not scoring on the index. The rating can rise above zero in cases of lucid or telepathic dreams (de Pablos, 2013). Whatever the rating, dreaming is an introduction to experiencing non-physical space, if one were to analyse the situation objectively.

The Index properly relates to separative experience, and to primary separation in particular. One may ask, why does the General Index not include the reported observation of physical things during a separative experience? It is commonly assumed that these reports rate the experience to be veridical, confirming its reality. But as already noted, this argument is weakened by the possibility that the information is generated by extrasensory perception, and is not necessarily by itself evidence of a claimed separation. As Whiteman stated, any apparent physical supporting evidence merely has "chiefly propaganda value for the uninformed or

Table 6-1 General Index of Reality

Factor	Rating Scheme
R. *Participated Reality, Grounding* Sense of ultimates, logical priority, life-fullness, substance, tangibility, vivid participation. **W**	No objective control or reflection. 0 Some dreamlike or imaginative quality, reflection and attempts at control. 1 More "real" than the physical, as regards participation and substance; free observation. 2 Strong participation, sense of substance and pin-pointed objectivity. 3
V. *Vertical Recollection, Valuing* Openness to higher direction, moral choice, guidance, and higher significance, Obedience. **S**	Complete lack. 0 Some awareness of moral choice or guidance. 1 Continuity of latent obedience. 2 Manifest obedience to the Source. 3
H. *Horizontal Recollection* Openness to rational continuity, memory of and comparisons with the physical state.	Severe lack. 0 Fairly good. 1 Complete openness to comparisons. 2
P. *Personal Communion, Purpose* Intercommunication of thought, feeling and impressions of character, openness to instruction, selfless and loving identification, sense of higher purpose. **E**	No person seen, or only dreamlike or "lay" figures. 0 Beginnings of objective knowledge of other minds, with communication of thought and feeling. 1 Clear and direct intelligible communication. 2 Communion in loving identification and interchange of wisdom. 3
I. *Integration, Release and Upliftment* Integration, joy, transformation; transcendence of emotionality, excitement and fear. **N**	Unpleasant or indifferent. 0 Beginnings of higher freedom. 1 A first integration and release from physical habits of thought; distinct improvement in bodily form. 2 Transformation to "proper form". 3
M. *Continuity of Memory*	No memory on return (only comes later). 0 No complete break, but some difficulty or vagueness of recall. 1 Memory continues vividly, as by perpetuation of the interior state, in detail without a break. 2

sceptical, who do not realise that separation is not established by them, but who may thereby be induced to accept the 'interior' testimony as having some bearing on 'scientific fact'."[86a,49]

This is not to deny that someone in a separated state may become indirectly aware of an event or feature in the physical world. The potentiality-field yielding a particular physical feature or event to a physical observer may yield a similar actualisation to a non-physical observer. This would account for non-physical observations being physically confirmed, but reliance still has to be placed on *internal* evidence that separation was experienced over and above any extrasensory perception.

If the evidence for separative experience has to rest solely on the declaration that the experience was felt to be powerfully actual, how convincing can it be? For Gotama Buddha the reality was considered to be unquestionable: after suitable mental preparation, one "produces disclosingly from this body another body, with bodily appearances made of mind, and with all limbs and lesser bodily parts, not deprived of any bodily function or sense. It is as if a man were to pull out a reed from its sheath. He would know, 'This is the reed, and this is the sheath'."[06,167] Whiteman held that the facts of separative experience "are as indubitable to me as is the truth of 2 + 3 = 5,"[94,84] (perhaps mixing logical and empirical truth).

The Index of Reality being based solely on internal evidence, Whiteman emphasised that the quality of reporting needs to be carefully scrutinised for it to be accepted as scientifically significant. Evaluation has to take into account whether the reporter shows understanding of features such as those included in the General Index, and is able to communicate this understanding in a convincing way along the lines of reliability and authenticity just discussed.

The Index might seem of limited value to standard parapsychology, where a useful scoring system would identify experiences that rate better than dream, hallucination or some form of pathology. In the Index the direction of ranking is towards mystical experience (items I and V), which is rarely looked for in parapsychology, even shied away from. The order of ranking is: undeveloped, borderline, psychical, pre-mystical, mystical. Out of a possible sixteen points, experience reaching six to eight points ranks as "psychical separations", scoring on free observation, awareness of substance and tangibility, openness to rational continuity and memory with a physical state. A further requirement of "mystical" rank is a "sense of ultimates" and "higher significance", "openness to guidance", "transformed being", and "manifest obedience to the Source".[86a,42;06,10] In the Index, items V and I single out mystical states and demarcate "mystical" from "psychical". Yet in the eyes of a physically-fixated investigator, the mystical requirements would indicate reduced reality rating, since "reality" means "like physical experience." For Whiteman this is a disastrous fallacy, centred on one-level naturalism. Highest reality may best mean "not like physical experience", not held captive in the powers of *māyā*.

This leads to the next question: if these experiences can be taken to be reports of something that exists at least momentarily, then what exactly is it? What kind of space is being talked about? The questions lead to such complexities that they deserve another chapter.

CHAPTER SEVEN

OTHER SPACES

In Chapter One it was noted that Whiteman wrote to Ian Stevenson, "everything hinges for me on the admission of 'other spaces'." What did he understand by "other spaces", or other worlds of conscious existence, and what led him to this understanding? Popular writing often assumes or implies that other worlds or planes or heavens have a permanent existence. Yet, as with the physical world, Whiteman pointed out that these manifestations are also products of the powers of *māyā* (Chapter Two), and we fall into what he termed "the fallacy of absolute objectivity" if we do not recognise this.[86a,183] He commented, "In other words, apart from the total circumstances on the occasion of observation, and the total potentialities behind, there is no coming-to-be in observable existence at any level."[86a,183] The "planes" of popular thought were seen to be characteristic of different kinds of mental state; no non-physical body or plane was believed to continue as an objective creation apart from its being observed: "only the unseen potentialities continue." (Letter to Scott Rogo, 15/6/1978). The matter of "other spaces" is anything but simple.

He was far from exceptional in finding there is something very wrong with current materialistic or physicalistic thinking about the experience of space. He was especially concerned that this thinking blocks an understanding of non-physical space encountered during separative experience. The error was seen to involve one of the powers of *māyā*: projection (Chapter Two), which leads to sensory fixation on what is projected, leading in turn to inferring a single space-time frame that is external to us, a single container of all that is projected, and which is taken to be the only "reality".[67,150;347]

The trouble with this one-level idea is that it makes it impossible to explain anything. As Heisenberg pointed out, explanation at the atomic or particle level rests on "a world of potentialities or possibilities rather than on things or facts." (1959, p. 160). The physicist David Bohm came to a similar conclusion that an "implicate order" (apparently close to potentiality or substructure) must be recognised, from which physically "an explicate order flows out of a law of the implicate order." (Bohm,

1980, p. 133). One-level naturalism cannot even explain how we think, since thinking requires at least some detachment or dissociation from the flow of physical time and events, discussed in Chapter Nine. As Whiteman pointed out, one-level thinking makes it impossible to explain consciousness, motivation, judgement, meaning, knowledge, morality—in other words, the essentials of being human. The only way out for someone immersed in the one-level idea is to deny the reality of these things, and especially deny reality to the more exceptional features of human existence, such as the occurrence of psychical phenomena, separative and near-death experiences, and mystical states.

Especially in separative experience, everything for Whiteman hinged on the admission of "spaces" beyond what is allowed in a one-level naturalism. Yet in most of the literature on separative experience, a one-level naturalism is assumed. In commonly reported cases of the experience, the person maintains that he or she "saw" their physical body from a separate position, usually above it. This implies that they were located in the same space as their physical body, which would be physical space.

This difficulty was first broached in Chapter Four with the dissected kidney analogy. We have to come back to the definition of "the physical". As discussed in Chapter Two, Whiteman applied the term "physical" to "kinds of observation, objects of attention or states of consciousness when the senses of the physical body are being normally used." In primary separation, which is the commonest state, there is no awareness in the physical body, and physical senses are inactive. If there were some detached visual apparatus floating in physical space it would be detected by physical means, and, quite apart from Kant's logical problem (Chapter Three), there is no firm evidence for this. By Whiteman's definition, then, in separative experience an observer and what is observed are not in physical space, even if a presentation is so similar to a physical scene that it is mistaken for it. He wrote, "If one is taken into a 'psychic' space when not familiar with such states, or with fixed ideas about them, and if the phenomena resemble physical ones closely (in technical language, the phenomena are *duplicate* ones), there may be a strong persuasion to think that the objects are being observed physically; and this applies even to possible duplicate presentations of the observer's physical body in its actual situation."[86a,11]

The term "duplicate" used by Whiteman may lead to misconceptions. One dictionary meaning of "duplicate" is: "existing in two examples", but this is evidently not how Whiteman intended to use the term. Consider the several reports of a patient claiming to have viewed an operation on his body while under anaesthetic. According to Whiteman's definition, the surgeons and equipment would be a non-physical presentation for the patient, since his physical senses are out of action. Does this imply the separate existence of a different, duplicate theatre and staff?

Not so, if one recalls the discussion in Chapter Two that a potentiality-field has the capacity to deliver different sets of physical and non-physical actualities, depending on the state of the observer. Someone who enters a separated state while physically unconscious will bring about an actualisation in the particular world his mind is functioning in, and it is quite likely that "fixed ideas" will tend to result in actualising a scene very much like a physical scene. This is because most people are accustomed to the ruling one-level idea, which assumes that only one "real" space can be manifested, namely the physical world. This fixed idea will contribute to the overall potentialities involved in a separative experience, and lead to the actualisation of a physical-like "duplicate" scene for the viewer, in this case an operating theatre. But there would be one single general potentiality-field, which would be actualised as a physical theatre to physical perception, and an approximately similar theatre to naive non-physical perception.

The more physically fixated the observer, the closer to a physical presentation the experience is likely to be; yet there is a gradation from experiences that closely duplicate a physical scene to those that show slight differences from the physical, to those in which a totally different manifestation is experienced. Slight differences from the physical are reported typically where the lighting differs from what would be expected in the current physical state—the scene is lighted at a time of physical night-time darkness; or part of a scene, such as around a bed, is brighter than the rest of the scene (Poynton, 2001). These small differences from the physical scene need not be unexpected, since different potentiality-actualisation processes must be involved in a separated state compared to a physical state.

This leads to research problems. Under the pervading one-level naturalism, any differences from a physical scene will be assumed by a physically-minded investigator to indicate error or hallucination, rather than that a different state will result in a different actualisation. Failure to recognise this tends to have a disastrous effect on out-of-body research. Firstly the experiencer will be inclined to leave out details of an experience that conflict with the current physical scene, to make it seem "realistic". This will tend to remove key data about actualisation in non-physical states. Secondly, investigators will be inclined to fall into the same error by accepting only physical-like reports. Critical data will then by thrown away in the belief that the reports are hallucinatory or "merely subjective".

Discernment of what is physical and what is a duplicate may become difficult when a separative experience is drawing to a close. Whiteman used the term "parallel world" for "a space very close to, or like, physical space, but not necessarily a duplicate one."[61,73] Here is a description.

> During this process, the inner space gradually melted away, at the same time as the space of a parallel world began to appear, both spaces being recognisable at once. Next, as the parallel world became predominant, the inward body began to be lowered into coincidence with (what appeared to be) the physical body, whose position was now clearly discerned. At the same time, the light of the other world began to close in, a boundary having been formed between it and the apparent light of the physical world, visible from the duplicate state. Then, lastly, as the arms seemed to come into coincidence with their physical counterparts, there came a flooding of life-energy, with a powerful and bitter tingling, entering at the solar-plexus and spreading through the body. With this, the connection between the spaces was complete, and consciousness appeared again in the physical world."[61,73;06,261]

Once someone becomes free of the constraints of one-level programming, Whiteman wrote of encountering an "inexhaustible variety" of experiences in other spaces beyond what is physically duplicated. Naturalistic conditioning may still result in "fantasy influences" and "fantasy forms or fantasy-like distortions of forms" in undeveloped states,[61,152] yet highest development leads to "perfectly acceptable forms of transcendent unified beauty," where "the surroundings, correspondingly, are distinguished by the quality of the light, and gain in *intelligible* character, perfection of beauty, depth of glowing heart quality, unitive freedom and sense of blending with other minds, as the highest condition, which is that of Mystical Form Liberation, is approached."[61,153] In these descriptions, Whiteman states, "'I' and 'me' do not stand for the familiar consciousness of self in the physical personality, or, in fact, for anything that can be known in an ordinary physical state of mind. To understand what is meant, the characteristics of the merged ordinary *self* must be ruled out."[61,158] There is then "recognition of *real being* behind the manifestations of 'personality'."[00,167]

These last three chapters on separative experience and other worlds deal with concepts of such complexity that it seems appropriate to review the most effective way of handling them scientifically. This is attempted in the next chapter.

CHAPTER EIGHT

SCIENTIFIC THINKING AND PRACTICE

Before retirement I ran a postgraduate course called Scientific Thinking. A good deal of it centred round the fact that most scientists generally don't think too much about the fundamentals of what they are doing. Received notions are stored in a back folder and one gets on with practical science. This point was made strongly by Whiteman, who recognised two modes of thinking in science. By far the most usual mode, which he called hypothetical-deductive, drew his criticism. The other mode, analytic-deductive, is rarely followed in general scientific practice, but was highly regarded by him as philosophical underpinning to theoretical physics, scientific mysticism, and psychical research.

The generally accepted way of doing science is that you come across a problem—some observation found by mistake or by searching—and you make a guess (dressed as a hypothesis) as to what it is, or how it works. You want to find out if your hypothesis is right, so you make deductions which you test for correctness. If the deductions fit what is observed, then you can take your guess to be reasonably correct, your hypothesis is confirmed. It doesn't matter how the guess was made; all that matters in science is that it is capable of being tested (Popper, 1972). You can never be certain your hypothesis might fail; new observations might show your deductions to be wrong, and so your theory needs scrapping or revising. There is no sure knowledge. The essence of the hypothetical-deductive mode is speculative theory-building developed as standard practice. Whiteman noted, "Those who are wedded to the scientific method called hypothetico-deductive, admitting no other, can become emotionally upset if that is questioned, even to the extent describable as 'frenzy' (of which I have had experience)."[06,21]

In contrast, in the analytic-deductive mode one is more interested in trying to apply conceptual analysis to a problem or observation. Take a simplified example: the observation that the sun and the full moon do not rise together. The hypothetical-deductive mode would make a theory that this happens regularly, and the theory is confirmed by finding that the sun and full moon have never been seen rising together. But you can never be certain your theory might fail, that an instance might turn up of the sun and the full moon rising together. On the other hand the analytic-deductive mode would analyse the geometry of the situation and conclude that it is logically impossible for the sun and full moon to be seen rising together as a natural event, because the sun can't light up the side of the moon facing away from it. One can be certain this is so; there is no unsure speculative theorising involved.

So if one is attracted to the analytic-deductive mode, one will look for the underlying logic of phenomena in the material world. Whiteman was confident that an "intelligible structure" can be discerned behind any particular observation, some normally-hidden substructure that is "operative in and analysable out of the total experience."[67,148] The finding of background logic in phenomena was, Whiteman pointed out, "what the great innovators in theoretical Physics were aiming at. And their chief results can in fact be derived by pure logic, apart from experiment usually suggested by it."[06,21] He cited Newton and Einstein among others. Thus, according to special relativity, "the rational structure of space and time and of the physical laws expressible in those terms cannot be comprised in a material world but belongs to a realm of *universal reason* (or general logic for all possible observers and points of view)."[77a,732]

This brought him in line with the German philosopher Edmund Husserl (e.g. 1931) writing mainly in the earlier part of the last century. Although not regarded as a mystic, Husserl's phenomenology struck a chord with Whiteman because of his description of a special kind of detachment that allowed intelligible analysis and immediate seeing in a state of "alienation from regular life".[67,404-409] This detachment, necessary in carrying out analytic-deductive tasks, had resonances with psychical and mystical experience. It insisted on the importance of direct experiencing free from the "cloak of ideas" that invests current theory-bound science. It chimes with Whiteman's definition of mysticism (Chapter One), "the facts and their relationship being known by the self-evidence of direct observation and not by reasoning or speculation".

The word "phenomenology" confusingly has two quite distinct meanings. Whiteman wrote that generally scientists "have come to use this word for 'looking into details of appearance', instead of 'penetrating behind appearances' to the structure of ideas which gives them meaning and substance."[77a,733]. The phrase, "penetrating behind appearances to the structure of ideas which gives them meaning", is a good portrayal of Husserl's phenomenology and of analytic-deductive science; but it tends to be pushed aside in favour of theorising in the hypothetical-deductive mode.

Husserl's phenomenology, of theory-cleansed and immediate seeing, clearly requires "a special kind of approach and a specially developed and open state of mind, if this 'immediate seeing' of intelligible characteristics and constitution is to be possible for us."[67,107] It involves "The process of 'holding time still' for the purpose of steady recognition of intelligible quality called 'stoppage'... By means of such 'stoppage of time' there dawns on us what Husserl calls 'pure consciousness'; and the faculty of discernment then exercised is called 'essential insight' (or insight into 'essences')".[00,88]

In scientific work, essential insight and analysis should provide "a clear-cut explanatory system ... as, for instance, the basic laws of mechanics are exhibited before any problems are tackled."[00,xix] Whiteman attempted to exhibit "basic laws" by formulating axioms or self-evident truths, because "if we can find concepts which are 'primordial' enough, all natural law could be developed from them by a kind of deduction."[77b,295]

An instance of a primordial concept was seen to be the hierarchical relationship between potentiality and actualisation. A simple example of this relationship is facial expression, where "the feelings of rage and fear obey one logic (as feelings dependent on thoughts and motives) and the shaping of the contours of the face depends on another logic (of muscular action connecting parts of the human body), and the two logics (at different 'levels') are connected."[06,151] One logic is inside or behind the other. If this hierarchical thinking is followed through, then some final primary or inner core understanding may be arrived at. Whiteman attempted this in a sequence of axioms in parapsychology, in the belief that an analytic-deductive method could be applied to this field.

The first result was a set of axioms published in 1973.[73,355] Axiom 1 read, "There is a *potentiality-sphere* for events in the physical world, whether or not they are observed by any individual. An individual can participate in this, and it can be called his *physical-perception* sphere. Actualisation of physical phenomena for him results from integration of potentialities of that kind with his subjective potentialities."[73,355]

He did acknowledge that "It is not known precisely in what ways individual potentialities can combine with impersonally physical ones so as to effect physical changes."[73,357] This is because "actualisation is the outcome of an operation which converts the unreachable into the reachable."[75a,199] He makes no bones about this being a very, very complex matter.

Another set of axioms was published in 1975 in a symposium on quantum physics and parapsychology.[75a,188] It starts from the grounds of "essence" and "existence", two terms related to potentiality and actualisation, which were given increasing importance in his work (Chapter Ten). An explanation was: "the *actual* flows with time, and can be said to be made into 'life' as by a spotlight passing over the total realm of *potentiality,* thus converting 'essence' into 'existence'."[93,17] The first of the three 1975 axioms concern potentiality, "unlimited in its variety and productiveness"; this includes consciousness, which is necessarily involved with actualisation at some particular time for some particular person. Treating consciousness as an aspect of potentiality formed the basis for his unification of psychology and physics, a topic explored in Chapter Ten.

The axiomatic method was not followed in later publications, although he continued to emphasise the need to uncover "a continuing *logical* and *creative* structure which can count as the origin of *all* appearances from *all* viewpoints and which is thus *not* 'contained' in physical space as a single appearance seems to be. Only this transcending logical structure merits the term 'reality', as opposed to appearance."[86a,140]

Whiteman, as with Husserl, held such logical structure to be open to comprehension or "essential insight"—or not quite entirely open, one might say, taking into account "unreachable" potentiality just noted. He described disclosure through essential insight to be "intellectual or perceptual knowledge which transcends in a certain clear and unmistakable way the onward urge of time, the rigid apartness of spatial objects, and the apparent isolation of the individual mind in its state of fixation on bodily impressions. It is a state of *release.* Fixation being overcome, the mind opens out into universality."[67,395] This leads to the question of what exactly is this "state of release", and how does one come by it? The question sounds germane to mysticism, and its huge size deserves another chapter.

CHAPTER NINE

SPACE, TIME AND RELEASE

To gain some understanding of the mind opening out to universality, we need to shed what Husserl called the "natural attitude" that hinders essential insight under a "cloak of ideas". Then, as Whiteman summarised it, "by a suspension of attachments, and a 'stoppage' (epoché), we come to a 'unique form of consciousness', characterised by 'a certain refraining from judgment which is compatible with the unshakable because self-evidencing conviction of truth.'"[67,108] (See Husserl, 1931 p. 109). It is an "operation which renders 'pure' consciousness accessible to us";[67,406] Husserl's declared "full insight" was that "Consciousness in itself has a being of its own ... which is in principle unique, and can become in fact the field of a new science—the science of Phenomenology." (1931, p. 113).

Whiteman observed that Husserl's descriptions of "stoppage" bear close resemblances to the Yoga Sutras (Chapter Twenty-one) and to early Buddhism, "a system whose entire teaching centres upon grades of 'release' and their attainment."[67,108] Sutras cited in Chapter Thirteen provide, in his view, "the cardinal principle of release,"[93,274] as did the Buddhist *jhānas* (*dhyānas* in Sanskrit), "explorations of the intelligible."

Whiteman summarised the *jhānas* as four grades or stages.[67,109;00,89;06,86] (1) Investigation, with hypotheses and reasoning. (2) Phenomenological analysis, revealing underlying intelligible structure directly, without theorising or reasoning. (3) Pure consciousness, focusing on timeless ideas while released from the body. (4) Transcendence of time, space and bodily impressions; free contemplation of everything, even our bodies, as a structure of "pure ideas". The *jhānas* were nevertheless regarded as being "pre-mystical" modes of insight of "this world-order", experienced as from the physical body, even though positioned on the "mystical way".[93,61]

Release from time-flow, as discussed in Chapter Three, indicates a state that passes beyond normal physical to some degree of dissociation.Chapter Seven discussed space in relation to scientific mysticism; some attention now has to be given to time. The awareness of time, or the limitations of time on awareness, has led to deep thinking ranging from science across to the humanities. T.S. Eliot (1969) wrote in his *Four Quartets* that to be conscious is not to be in time. Whiteman wrote that reflective thought is not possible without at least some detachment from the flow of physical time: "without reflection there can be no intelligent control, in fact no continuing recognition of objects and events; for reflection implies the ability to 'go back' in time."[00,88]

He called this process "time-release", "the process of 'holding time still' for the purpose of steady recognition of intelligible quality."[00,84] Marshall (2015) noted that "temporal cessation", where time or the sense of it seems to stop, is a feature of mystical experience. This requires the mental powers of "skills of intelligible time,"[00,85] namely stoppage or checking in Buddhism, the Yoga Sutras, and Husserl's *epoché.* As described in the *jhānas,* there are different grades of "time-release", from ordinary detached reflection to "a stage of release where we are fully opened to inspiration, with a perfection of judgement and choice which we could never achieve by discursive reasoning or deliberate directing of thought."[00,93]

The experience of becoming detached from the onward flow of physical clock-time is particularly familiar to musicians. Whiteman described it as "the faculty of analysing a heard melody or spoken sentence, when the whole is kept in mind beyond the flow of time, while the meaningful parts are distinguished pin-pointedly—time has become like space."[00,171] He cited Mozart's statement where a composition, "though it be long, stands almost complete and finished in my mind, so that I can survey it, like a fine picture or a beautiful statue, at a glance. Nor do I hear in my imagination the parts successively, but I hear them, as it were, all at once."[00,86]

Mozart's description suggests a kind of static time, or time becoming space-like so one can scan it backwards and forwards. It does not have the movement of clock-time, is not part of the physical world and can be accessed only through some degree of detachment from it, as in Mozart's description of being "as it were completely myself, entirely alone and of good cheer."[00,86] As Wagner put in the mouth of one of his characters in *Parsifal,* "here time turns into space." The experience is in the realm of "intelligible time",[00,85] when each detail or whole of a scene "is seen or recalled, as if *time stops;* and one is then able to go backward or forward from one thing to the next, freely as if one were contemplating a map."[00,149]

But, as Whiteman pointed out, "some kinds of insight in regard to time, very similar to those obtainable in higher dissociated states, can be experienced, without intelligent control, in states set up by drug influence."[00,84] He cited reports of experience under the influence of mescalin or LSD which feature a feeling of timelessness. These "lower" dissociated states are without intelligent control.

Whatever the cause of time-transcendence or feeling of timelessness, it seems that two time-levels may be experienced, one having physical clock-time progression, the other not progressing and space-like. The progression of physical time may be seen causally linked to time-ordered potentiality-fields, something like features on a spatial map. These features do not themselves progress any more than roads marked on a map progress, yet they have the potential to result in a succession of events actualised in the physical world. Looked at this way, Whiteman noted, the potentiality of a past physical event could still be actualisable in an individual's thought-image sphere, although irretrievable in the physical world. Likewise, a future physical event could be actualised in this way, resulting in a precognitive experience. Yet, as noted in Chapter Two, the time-ordered level of potentiality is not rigidly fixed in the way a map is. It is accessible in states of physical dissociation, and so allows alteration of time-ordered potentiality and the exercise of free will.

The process of actualisation itself needs to be considered. Granted two levels of time, static and moving, we need to account for "a passage from the intelligible exactitude present in the physical potentiality (e.g. the structures and laws of Euclidean geometry) to the inexactitude that is inevitable in all physical measurement or personal estimates of measure."[00,95] A third factor of time is therefore needed. As a result Whiteman recognised three very different "time-like functions": intelligible or potential structuring signified by T; the process leading to physical actualisation, to change, symbolised by τ, and an instant of measurable physical time, denoted t.[00,95] Only isometric clock-time t is measurable, and the combination of time t with the three spatial dimensions, length, breadth and height provides the measurable framework for ordinary life. The combination of the three spatial dimensions with all three time dimensions provides a means of accounting for very much more, as will be discussed in the next chapter on physics and psychology.

One could describe non-physical and mystical experience as freedom from physical space and time, but not necessarily dispensing with space and time altogether, for spatial features and time can be actualised in non-physical levels. Whiteman was emphatic that spatial dimensions and relationships are encountered in non-physical experience: "let no-one suppose that the term 'space' is wrongly applied to the non-physical, for in all kinds of non-physical sensing, objects have extension, position, direction and shape all rationally structured as in physical space, and capable of being moved about in that space relative to other objects there."[86a,6] They are "subject to the same geometrical laws as hold in the physical world."[77b,302] Yet there may be a lifting of limitations that are encountered in physical space, such as freer movement in a non-physical space, giving the sense of release.

As noted in Chapter Three, a blanket use of the term "non-local" in place of "non-physical" does not take into account the local placing of objects in experienced non-physical space. "Non-locality" in Whiteman's usage applied only to atomic levels, discussed in Chapter Three: "we must not picture events there as if in normal physical space-time. Today, they are commonly spoken of as subject to 'non-locality'."[06,326;Appendix A] But these pictured events bear no relation to the experience of located objects in non-physical space, decidedly "local" to the experiencer.

Consciousness was seen to be "unlocated".[77a,744]

Whiteman placed greater emphasis on release from physical time rather than physical space. Husserl's idea of stoppage of time is crucial, resulting in a "unique form of consciousness." The state of time-release is the result of practice, summarised by the *jhānas* which "develop successively more far-reaching controls of intelligible time." The second *jhāna* was described as Active Recollection. Whiteman used the term Recollection in a sense used by some Christian mystics, which was "authoritatively declared to be the foundation of all voluntary practice leading to mystical skills and knowledge."[86a,18] Its opposite was regarded as psychological fixation, and giving an analogy from chemistry,

> *fixation* is the process of combining a gas with a solid, and naturally getting something with quite different properties. If that earthy solid is then somehow decomposed, we might re-collect the gas in its pure form. In the same way, we may imagine that the pure and freely-flowing spirit is combined with something restrictive so as to make the crude sensations of this physical world, subject to the urge of time. But by some carefully-guided operation it might be possible to restore freedom to the spirit and re-collect it in its proper sphere.[61,8]

He listed no less than twelve "practices or principles" for "recollecting our own supreme wakefulness".[61,9] Recollection could be said to be broadly comparable with mindfulness, but more penetrating.

Active Recollection was a discipline developed first by focus on physical sense-perceptions. "In every case the aim must be to recall the *conceptual essence* of what is perceived, so that eventually everything can be pinpointedly recalled in a releasing way, without any feeling of obstruction."[86a,19] He found a Pelman course in mind training helpful in developing "this power of timeless recall",[00,315] described in Chapter Thirty-one. Active Recollection, then, is "the cultivated power to withstand and surmount tendencies to automaticity, absent-mindedness, self-absorption and indulgence",[86a,18] or "focus, plus letting-go, and passage to the 'time-stopped' purely conceptual (as distinct from the supposedly mindless objectivities of the worldly life)."[86b,642;00,171]

The third *jhāna* was described as Continuous Recollection, which "presents itself, after extensive practice of Active Recollection, as the *freely stabilised ground of release* at which Active Recollection has been aiming. It is the 'pure consciousness' which is made accessible to us by that first practice of control and which opens us to what Husserl calls 'primordial object-giving'."[86a,19] It is "letting-go, plus expansion, and effortless removal of conditioning (fixed habits of thought, etc.); stabilizing of the ground for unconditioned observation at all levels of life."[86b,642;00,171]

It may be noted that this practice of Recollection does not involve long periods of meditation; it is, as Whiteman often remarked, a state of being in the world but not of it. Nevertheless, while he stressed "the importance of Continuous Recollection for a spiritually healthy life in this world;" it is only "the threshold of the Mystical Life, not the final attainment."[93,294]

This leads into an area that was of greatest importance to Whiteman: the union of physics and psychology. There are many complexities here, which deserve a chapter on their own.

CHAPTER TEN

PHYSICS AND PSYCHOLOGY

The idea in quantum theory of the "indivisibility of the phenomenon" of viewer and viewed was discussed in Chapter Two. As Whiteman put it, "nothing specific can be actualised without the integration of 'subjective' and 'objective' factors, not in succession, but on the instant of actualisation."[86a,240] Integration involves a joint actualisation of the "impersonal" potentialities of the event and the "personal" potentialities of the observer, an integration of physics and psychology. The need to clarify this integration becomes ever greater as the study of physics and of consciousness progresses towards a "reconciliation of science and spirituality" (Kelly *et al.,* 2015). Carr (2015, p. 269) argued that as mental phenomena involve some form of space, and as aspects of the physical world require hyperspatial treatment, "it seems natural to relate these two ideas" as a hierarchical "higher-dimensional psychophysical information space."

In agreement with this concept, Whiteman used as his starting-point a trend led by Galileo in seventeenth century science, one that drew a practical distinction between impersonal and personal phenomena. Galileo applied the term primary and secondary qualities respectively. The primary-secondary distinction is, essentially, between physical, impersonal conditions and one's personal experience of them. Physics is conventionally concerned solely with primary qualities: position, motion, mass. From the seventeenth century onwards these were seen to be measureable—at least approximately—and so invited scientific investigation. Secondary qualities, which seemed unreceptive to measurement, were initially of no concern to the brave new science, even though to ignore them meant passing over what is actually experienced. Early attempts to investigate the secondary qualities of mentality were largely mechanistic, taking their cue from classical physics as being proper "science". In materialistic psychology there are still attempts to replace secondary qualities by primary equivalents; physical signals that allow measurement then become the sole Measure of Man. This is all that can be attained in a one-level naturalism.

Examining a possible link between primary-quality physics and secondary-quality psychology is certainly no simple matter. For a start, has psychology been systematised to the same degree as physics to allow a close comparison of concepts? The answer broadly must be, no. The task, then, as Whiteman saw it, was to uncover a system in mental functioning as precisely analysable as phenomena studied in physics, and show parallels between the two while not losing a distinction between primary and secondary qualities. The way of doing this was to develop number systems from first principles, such as the axioms discussed in Chapter Eight; these would give a handle on making comparisons, and so bring understanding to the integral relationship between viewer and viewed.

For psychology, he saw that "the 'number systems' in Eastern mystical philosophy, especially the Upanishads (with some confirmation by Western psychology and direct mystical testimonies), provide a basis for precision, in such close correspondence with the formulations of quantum theory that there is an obvious invitation to exhibit the two as aspects of an integrated understanding of the world."[77a,741]

Number systems, as he pointed out in several places,[86a,192] conform to groups and cycles both in physics and psychology. In physics there are, for example, the four dimensions of space-time, and in quantum field theory the 16 components of the gravitational and energy-momentum tensors. In psychology he placed importance on a cycle of four "functions" in creativity, recognised by Poincaré, Freud, Jung and others.[77b,300;86a,193] This four-fold process occurs in the learning cycle: "purposive *drive,* deciding on *means,* putting into *practice,* and a virtually secret *maturing* of one's skills in consequence."[06,314] He pointed out that this four-fold cycle of functions was evidently recognised throughout the ancient world, most commonly represented as four stages or directions in the passage of the sun: rising from darkness in the east, to zenith in the south (in the northern hemisphere), down to earth in the west, and underground as incubation in the north.[06,314] This corresponds with aspiration, assessment, action, fulfilment; or more broadly:

east – illumination, intention or purpose, drive, enthusiasm
south – response, valuation, means, judgement
west – application, action
north – release, non-attachment, incubation, latent in the start of a new cycle.[86a,193; 06,68,119]

Whiteman placed great importance on the cycle having its "counterfeits" of self-will (E), self-satisfaction (S), automaticity (W), complacency (N), leading to a corresponding cycle of "stresses" of death-feeling (E), shame (S), pain (W), fear (N), which in turn may lead to "inner contests" to bring about "right value-building".[00,31] These inner contests are "set up between the unwise impulsions of self and the inklings of wisdom."[00,117] In what he regarded as "lower levels" of spirituality, the sequence of a creative cycle tends to be scrambled. Then, "the outgoing third function (observed action) may seem to be first in creative order, while self-will (counterfeit of the first function) may seem to be a resulting (mentally) active one; and similarly the second function (counterfeited as self-satisfaction) may seem the final goal." So "materialists will naturally think that the 'down to earth' actions in the world come first, and self-awareness is a consequence."[00,65] Active Recollection connecting the third and fourth functions breaks this fixated or lower cycle to a free or released cycle.

Table 6-1 reproduced in Chapter Six integrates the meaning of the four functions or directions with factors of the General Index of Reality. E was linked to P, purpose and personal communion; S with V, valuing and vertical recollection; W with R, grounding and participated reality; and N with I, release and upliftment, integration.[06,10]

Whiteman noted that three of the stages, East, South and North are composite, containing the two "first principles" of essence and existence, which were touched on in Chapter Eight as the broader view of potentiality and actualisation. In turn, essence and existence each have three constituents, termed by him "hypostases", meaning underlying, irreducible causes that are not attributes of something else. The three hypostases of essence were seen as purpose, means, end. The hypostases of existence were actuation, becoming, manifestation.

Principles	*Hypostases*
Essence	Purpose
	Means
	Ends
Existence	Actuation
	Becoming
	Manifestation or grounding

The three hypostases of essence involve causation at the level of timeless potentiality, while the hypostases of existence are "quasitemporal causations",[77b,299] the "three causal steps in the actualising of potentiality"[00,98], called in Upanishadic philosophy the Great Stride of Vishnu (a principal Hindu deity, the Omnipotent, cited in Chapter Twenty-two).

> Manifested, constituted, secretly proceeding,
> this is the "Great Stride".
> So, this universe is caused to come forth.[06,149]

As an example of how this system applies to mental secondary qualities, "deciding on means" (south quarter) combines the timeless concept of "means" (essence) firstly with "actuation" (existence), then with "becoming" (existence), and lastly with manifestation (existence). This analysis allows the development of a number system in psychology, and so the prospect of comparison with a number system in physics.[06,314] The hypostases of essence were numbered 1 = purpose, 2 = means, 3 = ends. The hypostases of existence were numbered 6 = actuation, 5 = becoming, 4 = manifestation/grounding (the reverse numerical order was to facilitate comparison with physics, to be discussed later). So "East" can be expressed as the combination 16 (enthusiasm), 15 (inspiration), 14 (intention); "South" as 26 (obedient response), 25 (precising), 24 (judgement & choice); "North" as 36 (non-attachment), 35 (recollection), 34 (expansion into wholeness). The third quarter, West, lacks the "essence" hypostases, since it brings prior causes into manifestation purely as "existence", the motivation and means being already decided. It therefore comprises only 46 (actuation of grounding), 45 (becoming of grounding, application), and the actual manifestation or actualising itself. As this is not a causal phase, it can be given no number-notation.

This leaves the combinations 12, 13, 23 and 65. Whiteman regarded these as having a linking character. For instance, 13 links responsive valuation (S) to consideration of purpose (1) and end (3), and 65 links the actuality of grounding (W) to consideration of actuation (6) and becoming (5), allowing a sense of fulfilment of what was intended and decided on.[06,315]

The twofold combinations of the numbers 1 to 6 give a total of 15, four in each quarter except W, which has only three. These 15 combinations were termed "subfunctions" since they make up the four "functions", an equivalent term for the four "quarters". Addition of the numberless "44" to the West function raises the number of subfunctions to 16, "44" being "the final unnumbered creative result" or "fulfilment in action", in agreement with "15 + 1 *kalās* ('parts') in any creative act", as recognised in the Upanishads.[06,71;315] The sixteen-fold cycle is described, for example, in

Whiteman's translation of the Bṛhad-āraṇyaka Upanishad. The "break of day" refers to fulfilment or actualisation.

> The Creator-Lord is this cycle with sixteen stages (*kalās*). His "nights" are just fifteen stages, the sixteenth being a settled stage. ... Having in the sixteenth stage entered everything carried by the spirit he is born thence at break of day.[06,71]

The functions and subfunctions are tabulated below. Whiteman's remark may be noted, that "The names of the various hypostases and subfunctions are suggestive merely. To arrive at their intended meaning, it is necessary to stabilise the relevant function operating creatively in one's own mind, and to see that each subfunction involves the stated hypostasis."[86a,204] As a result, the names applied to the hypostases and functions/subfunctions vary from place to place in his work, depending on the special meaning in that context. He saw the subfunctions to be building blocks of an individual's functioning, operating in cycles continuously and in any world. They arise from the general "structures of reason which seem to be primordially involved in psychology."[77b,297] Yet beyond that was the attempt to match this 16-fold system with 16-fold systems in physics, and so bring about a union of physics and psychology. To achieve this, what may be called the "subjective" subfunctions just considered (Galileo's secondary qualities) need to be matched with "objective" combinations of space and time derived from physics (Galileo's primary qualities).

The previous chapter discussed Whiteman's idea of three time-like dimensions or variables, T, τ and t. Time can flow (t), be stopped as in Huserl's *epoché* (τ) or be spread, as detected in precognition (T).[06, 317] If these are combined with the three spatial dimensions of length, breadth and height, *x, y* and *z* respectively, a six-dimensional system results. As outlined in Appendix A, they were combined using a six-dimensional tensor calculus and sedenion algebra,[77a;77b;06,322] but taking a simpler line, if the six components of space and time are paired, the result is 15 combinations. Added to this is a scalar term bringing the number to 16, as in the subjective cycle.

Length, breadth and height were allocated the numbers 1, 2 and 3, "agreeing with accepted terminology in physics."[77b,298] To combine this with psychology, a link between the physical dimensions and the number-system in psychology had to be established. This was said to rest on a "fundamental mystical teaching" that three "quasi-spatial concepts" (before and behind, right and left, height and depth) "correspond to the three categories of essence, when these characterize an individual state of consciousness in regard to wisdom."[77b,298] The three hypostases of essence—purpose, means, and end—then correspond with the dimensions length, breadth and height, both numbered 1, 2, 3.

The three time-like dimensions T, τ and t correspond similarly with the three hypostases of existence: actuation or impulsion, becoming or change and flow, and grounding or actualisation. They were numbered 6 or 0, 5 and 4 respectively.[06,322] T is 0 because "There does not seem to be any one right way to allocate values to the 'planning' or 'cosmic' time variable T."[77b,306]

"Everything knowable," he wrote, "is, in essence, a 16-fold structure of concepts associated with an identifiable 'source of consciousness' (in itself unlocated). ... At each level where 16-fold cycles project a 'world,' the possibility of consciousness requires also the corresponding manifestation of 16-fold cycles of psychological action, finally expressing themselves correspondingly in bodily action at the physical level."[77a,744] To complete the union of psychology with 16-fold physical objectivity, therefore, "one needs to exhibit 16 psychological categories in every phase of consciousness, coordinated in a cycle which exactly corresponds to the 'objective' cycle or set of 16 σ-operators by means of which a 'world' is projected."[77a,745] This is best shown in a table that lays out the relationship between the "objective" combinations of space and time, and the "subjective" subfunctions.

Table 10-1 Objective and Subjective Cycles

Quarter (function)	Objective combination	Space/time symbols	Subjective subfunction	Psychological term
E	10	*xT*	16	Enthusiasm, impulsion of purposes
E	15	*xt*	15	Inspiration, flow of purposes
E	14	*xτ*	14	Intention, actualisation of purposes
E	23	*yz*	23	Consideration of fulfilment
S	20	*yT*	26	Response, impulsion of means
S	25	*yt*	25	Precising, flow of means
S	24	*yτ*	24	Judgement, actualisation of means
S	13	*xz*	13	Consideration of purpose
W	40	*τT*	46	Extroversion, impulsion to ground
W	45	*τt*	45	Application, movement to ground
W	16	—	—	Actualisation, identification
W	50	*Tt*	65	Consideration of actuation
N	30	*zT*	36	Non-attachment, release
N	35	*zt*	35	Recollection
N	34	*zτ*	34	Expansion
N	12	*xy*	12	Consideration of means

Whiteman never published this complete table, only fragments in a number of writings.[77a;77b;86a,06] The rationale behind this scheme was presented also in a rather fragmentary way, which makes it especially difficult to follow. Essentially, he adopted the analytic-deductive method (Chapter Eight) in which "by a logical argument concerning the most archetypal categories of existence we may arrive first at a 16-fold cycle of potentially *conscious* functions (the 16 *kalās* of the Upanishads), and then find this in correspondence with the 16-fold cycle of elementary particle physics, concerned with objective spatio-temporal manifestation."[86a,191] The subjective cycle "parallels exactly what may be called the 'world cycle' (as it appears in Quantum Field Theory) in every way, while manifesting itself in every kind of human experience. This is a form of *potentiality* which is discernible, and indeed can be *lived* by the individual in an enormous variety of altered states of consciousness."[06,332]

This is the core of the matter, first broached in Chapter Two, where it was noted that the demand of quantum theory, as Whiteman saw it, is that "the 'purely physical' is incomplete, and manifestation requires that the circumstances and mental powers of an individual observer can be given before anything definite can be manifested."[00,132] Nothing definite is actualised in the objective cycle of space, time, mass, "unless there is combination with a subjective cycle determining the observer's response."[06,316]

A description of these structures and correspondences is (if accurate) simply describing a part of "what is"; "the inner constitution of nature," as his *Philosophy of Space and Time* was titled. Yet it was never intended to be a mere academic, abstract numbers game. Implications and uses will emerge in later chapters of this book, as in Chapter Twelve, which shows how understanding the functions is important in questions of responsibility and criminality, and Chapter Twenty-four, examining Isaiah's metaphors of mystical development.

Whiteman was at pains to show that the essentials of these ideas reach back over thousands of years to the Upanishads, the Buddhist canons, Yoga Sutras and Plato. His treatment could be regarded as updating ancient wisdom in this area. He cited three Upanishads in relation to the sixteen *kalās,* described as the "operational parts in creation".[86a,201] He emphasised that the *kalās* were expounded "in terms of direct experience and not philosophical speculation."[93,73] To illustrate the importance placed on the *kalās,* the Bṛhad-āraṇyaka Upanishad was cited: "A man who knows this is himself that conjunct proceeding, that Ruler of Creation with 16 factors."[77b,187]

Western conventional thought being what it is, all of this tends to make few inroads into science and academia. This applies also to the rest of the kinds of experience and analysis discussed in previous chapters. This Part on the basics of scientific mysticism therefore has to conclude with an enquiry into the indifference and incredulity that pervades the westernised world.

CHAPTER ELEVEN

THE IRON RULE AND IMMANUEL KANT

> "I say it is a scandal that the dispute as to the reality of [non-physical] phenomena should still be going on, that so many competent witnesses should have declared their belief in them, that so many others should be profoundly interested in having the question determined, and yet the educated world, as a body, should still be simply in the attitude of incredulity."

This could have been written today, but actually it was part of an address given in 1882. The speaker was a Cambridge professor, the occasion was the first presidential address to the newly-formed Society for Psychical Research in London (Sidgwick, 1883). The speaker, Henry Sidgwick, added, "scientific incredulity has been so long in growing, and has so many and so strong roots, that we shall only kill it, if we are able to kill it at all as regards any of those questions, by burying it alive under a heap of facts." (p. 8.) We should "trust to the mass of evidence for conviction." (p. 12.)

The "scandal" continues, even though the heap of facts has become a solid mountain (e.g. Kelly *et al.*, 2007; Broderick & Goertzel, 2015). Sidgwick evidently miscalculated the strength of the many and strong roots of incredulity. Our culture is firmly rooted in it, and any counter-culture that incorporates Whiteman's mysticism remains on the fringes of the "educated world", whatever the mass of evidence.

How did this come about? "Scientific incredulity", after all, is not a feature of every world culture; it is a special feature of relatively recent Western-based culture, although now spread across the globe. A materialistic philosophy can be traced back to classical Greece, and remained an underlying current until its burgeoning in the Enlightenment with "its rejection of mysticism in favour of reason." (Vitzthum, 1995, p. 9). Historians usually trace this burgeoning back to the seventeenth century with the activity of Galileo, Kepler, Newton, Harvey and many others. As noted in the previous chapter, their methods of investigation resulted in a mechanistic view of nature. It was useful for carrying out investigations: reduce phenomena to manageable fragments that yield to measurement. But viewing nature as composed of measurable bits and pieces leads to the thought that this is all there is to nature. Any conceivable interaction must be like clockwork, completely determined. In a science focused on what is physically measurable in some way, there was no place left for any idea of spirit. The fragmented thinking denied wholeness in nature. Jan Smuts, in his seminal *Holism and Evolution* (1926, p. 99), saw this as "the iron rule of the mechanistic regime".

A highly influential advocate of this iron rule was the philosopher Immanuel Kant, writing towards the end of the eighteenth century. Kant's lofty position still casts a long shadow over psychical and mystical research, throwing it out of court or at least relegating it to the academic fringe. His position therefore needs investigating.

Essentially he maintained that you cannot experience anything beyond what is physically presented. Whiteman was in agreement that ordinary practical experiencing (*Erfahrung*) and ordinary understanding (*Verstand)* will not directly disclose non-physical levels and intelligible structure, but he pointed to a third sphere of experience where this can succeed.[67,394] Long established in Platonic and Eastern traditions, it was not favoured by the "Enlightenment" thinking that Kant was settled in. The third sphere of experience reaches beyond *Erfahrung* and *Verstand* to a "centrally creative and causal category of knowledge", one that reaches beyond what Kant recognised as "ordinary practical experiencing and understanding." It involves (to quote again) "intellectual or perceptual knowledge which transcends in a certain clear and unmistakable way the onward urge of time, the rigid apartness of spatial objects, and the apparent isolation of the individual mind in its state of fixation on bodily impressions. It is a state of *release.* Fixation being overcome, the mind opens out into universality."[67,395]

The varieties of psychical and mystical evidence, Whiteman argued, reveal this "third class of experience" which is "neither physical nor subjective."[61,114] It can be classed as "objective non-physical experience" because, as in physical life, there can be full powers of detached observation, reflection and rational judgement, together with a large corroborating literature from many parts of the world, indicated in Part 3 of this book. In such states of "unified detachment" (discussed under phenomenology in Chapter Eight) "we may expect to learn, by the clarity of their intelligible representation, the truths concerning the structure of personality, conditions in 'other worlds', and principles of hierarchical causation between one sphere and another."[61,114]

Whiteman noted that Kant's position made him deny the possibility of telepathy and precognition; he cited a passage in the *Critique of Pure*

Reason of 1781: "a special ultimate power of observing the future in advance (and not merely inferring it), or lastly a power of standing in community of thought with other men, however distant they may be—are concepts the possibility of which is altogether groundless, as they cannot be based on experience and its known laws."[67,394] This view is still widespread, yet one has to ask, did Kant have any good reason or evidence (or even understanding) for declaring this?

Clues are given in an invaluable collection of Kant's minor writings on this topic edited by Gregory Johnson (2002). It contains Kant's early work entitled *Dreams of a Spirit-Seer* (1766), which is a debunking attack on the scientist-turned-seer Emanuel Swedenborg, an early contemporary of Kant. Swedenborg carefully described meetings with spirit people, stating that his accounts were *ex auditis et visis,* from seeing and hearing. Yet Kant declared, "All spirits and ghosts, apparitions, dream interpretations, precognitions of the future, sympathy of souls are altogether a most objectionable delusion, for it does not allow itself to be explained through any rule or through comparative observations ... and even if real ghosts exist, a rational person must still not believe in them, because it corrupts all use of reason." (*Volckmann Metaphysics c.* 1784, in Johnson, 2002, p. 97).

As with the statement from the *Critique,* this declares that topics such as covered in scientific mysticism and psychical research are an "objectionable delusion" because they are outside the bounds of proper observation. If one adheres to a one-level naturalism, then observation must necessarily be limited to the physical world, but Kant gave no substantial grounds for rejecting Swedenborg's claim that his accounts were "from seeing and hearing" *outside* the bounds of physical observation, the "third class of experience" noted by Whiteman. In Kant's statement about ghosts it appears that observation is held subservient to the "use of reason"; but if "reason" is placed above observation and experience then what is there to distinguish it from prejudice and dogmatism?

One may ask, was a philosopher of such great renown really inclined to engage in prejudicial thinking? The answer is given in his naming (in 1775, for the first time) the different "races" of humanity defined by skin colour and place of origin (Jablonski, 2012). For him, the races were fixed and immutable, arranged in a hierarchy from Europeans at the top through Negroes to American Indians at the bottom. We have rejected this one-time influential prejudicial racism, yet his prejudicial stance on psychical phenomena is still nearly as strong as ever. Kant caustically dismissed well-documented reports of Swedenborg's psychic ability; his prejudice is revealed in a letter to the philosopher Moses Mendelssohn, which stated that on reading Swedenborg, "I saw that I would never have peace from the ceaseless questions of those who supposed they knew something about this until I had disposed of all these anecdotes." (Letter of 1766.) One still sees strenuous efforts to "dispose" of such troubling reports, with policing powers advocated by Kant: "science arrives at the determination of the limits set for it by the nature of human reason ... then the boundaries draw closer together and marker stones are laid that never again allow investigation to wander beyond its proper district." (*Dreams of a Spirit-Seer,* ed. Johnson, 2002, p. 60.)

This policing of investigation beyond a "proper district" is still much in evidence today, and shows Sidgwick's hope of burying the attitude of incredulity under a heap of facts to be over-optimistic. It was an Enlightenment ideal to glorify "human reason", but it can become a cramping application of Smuts's iron rule if free observation and experience become subservient to it.

It has to be said that Kant himself admitted in his letter to Mendelssohn that he was "conflicted" over the Swedenborg issue. In the document *Metaphysics L1 c.* 1779 (in Johnson, 2002, p. 89), not meant for publication, he even wrote, "The thought of Swedenborg on this matter is quite sublime." He speculated about "other worlds", in a way compatible with Swedenborg. He wrote (original italics), "the separation of the soul and the body consists in the *transformation of the sensuous intuition into spiritual intuition; and that is the other world.*" This seems to be an admission that it's possible to experience what is beyond physical presentation; also it can be read as conveying the idea of interpenetrating spaces discussed in Chapter Seven.

It can also be read as a split if not duplicitous mind. In contrast to some of his private writings, his published work casts the long shadow that dismayed Sidgwick, and which still lies over the fields of psychical research and mysticism. Kant was simply wrong in his public dogmatic declaration that these fields contain "concepts the possibility of which is altogether groundless, as they cannot be based on experience." There is now a massive literature based on empirical evidence (e.g. Kelly *et al.*, 2007; Kelly *et al.,* 2015; Broderick & Goertzel, 2015). Wrong also is his dismissive statement, "All spirit-apparitions are, of the kind that we can neither set up experiments nor precisely observe and inspect them, and it thus allows reason no further employment here at all." (*Metaphysics L2,* 1790, in Johnson, 2002, p. 101). This is now a field of empirical research with a large literature (e.g. Kelly *et al.* 2007). Yet the current practice of labelling such research "pseudoscience", noted in the Preface, is a continuation of Kant's attitude, an attitude that harbours bias and prejudice, a betrayal of ideals of Enlightenment.

Kant's attack on Swedenborg can be seen at least in part as a smear campaign that suited him; it helped to establish his position in academe as a critical thinker by associating himself with the sceptical tenets and attitudes of the times (Johnson, 2002, p. xvi). This strategy is still successfully followed to this day.

One may nevertheless ask, is there any way the public side of Kant's work can be reconciled with scientific mysticism and psychical research once his dogmatism and prejudice are put to one side? A key issue lies in his "rational" placing of limits on human experience and investigation, which was aimed to avoid "fruitless enquiry in respect to a question that requires data from a different world than the one he senses." (*Dreams of a Spirit-Seer* p. 57). He did not take account of disciplined practices even in his time that could change the limits of experience and permit "data from a different world"; these were largely Eastern and so would count for nothing among superior Europeans.

The situation today is very different. There is widespread dissemination of Eastern mysticism, mind-altering techniques of meditation, drug use, psychical research and exploration of consciousness generally. Kant had argued in his *Critique of Pure Reason* (1781) that the nature of our perceptual faculties determines how the world seems to us. Put drastically, what our state is, that is what we perceive. More recently it has come to be realised how radically one's state can be altered. Alteration of state will lead to an alteration of one's mode of perception and what is experienced (Chapters Two, Seven, Ten). So "data from a different world" may be expected by bringing about altered states of consciousness (Tart, 1975), states that open the gates of perception. Mysticism and psychical research, whose rigorous aim and practice is to open these gates, is therefore not necessarily in conflict with an extension of Kant's concept of how the mind works. Despite this, the iron rule that was to follow Kant's attitude of incredulity still remains an unjustifiable scandal, as Sidgwick observed.

The Preface to this book started by noting that many people see this century as bringing in a change to this position, a new Renaissance. In this renaissance, a science of mysticism has a potent place, which the chapters in Part One have attempted to outline. It is time now to discuss how scientific mysticism sheds light on the structure and workings of the individual person, and to explore the broad world-view that it can generate.

Part II

The House of Personality

CHAPTER TWELVE

THE MIND OF A MURDERER

In the Tibetan Wheel of Life the outer rim shows the unfoldment of *saṁsāra,* "the ephemeral world of birth and death" (Govinda, 1960). At its fifth interval is a house with six windows; it represents what Lama Govinda called the origination of a psycho-physical organism from the underlying samsaric nexus. The six windows represent six senses, which in Tibetan mysticism includes thinking. The house is a busy place. Gotama Buddha noted it was "Just as if there were a house with an upper terrace and a man standing thereon should watch men entering the house and coming forth out of it." Whiteman cited this analogy from the Buddhist Dīgha Nikāya.[86a,99] It continues, "Then he would know, 'Those men are entering the house, those are leaving it.'"

What he would know, Whiteman held, is that there is a continual coming-and-going of influences through "the house of the personality". Or, at least he would know it if he were alert and suitably detached from normal samsaric life. Most of us, most of the time, are not aware of what is coming and going, of influences good and bad coursing through our samsaric house. This situation was of deep concern to Whiteman in the case of two murders in his home city of Cape Town, which he examined in great detail. To him the question of what goes on in someone's house became pivotal: did the culprits recognise untoward influences entering their houses, and how did they and the courts deal with the situation?

It leads to the question, how stable and under control is the community within a house of personality? It is a question that has to pull together several threads in science, mysticism and psychical research.

In the Buddhist canon and the Yoga Sutras (Chapters Twenty-one and Twenty-two) the coming and going of influences in the personality "results in a succession of personality-phases, or Jungian *personas* in a very broad sense of the term."[93,53] The important term *jāti* is applied to these "births", which "can be very brief, and innumerable in the physical life; or they can be extensive personality changes or upheavals, as in the case of so-called 'multiple personality' or other kinds of mental disorder."[93,53]

In relation to the actualised physical behaviour of an individual, Whiteman saw personality to be in the realm of potentiality; the term "personality" was taken "to signify all that lies causally behind the behaviour of a certain physical body—provided something of the individual's memory is called upon in that behaviour." [00,103] The occupants of a house of personality should therefore be seen as purely causal, and not given any meaning of solidity. "Just as the phenomena studied in physics have an elaborate substructure of multi-dimensional state-functions not physically observable, so it has to be admitted that our conscious life at the physical level has an elaborate non-physical structure of motive forces, and potentialities for sensing, feeling, remembering, belief, and so on, all normally subconscious for the individual."[00,119] The case of the two murders is followed in the next seven chapters as a focal point to develop these ideas, before discussing release from the samsaric house.

An adolescent girl, Marlene, was said to have been "of quiet and very scholarly disposition, with a notably religion-oriented upbringing, and on all accounts deeply moral and gentle in nature." She was even described as being "mystical".[00,344] In 1971 at age sixteen she left school as a model pupil (she won several cups and certificates for music and Bible study), and took employment in an orthopaedic workshop. Around age eighteen, after a disrupting scene with her father, she left home, abandoned her homely nature, began to socialise and respond to men's interest. It led to starting an affair with a fellow employee, married and old enough to be her father, as sadly happens with adolescent girls. She fell intensely in love, but he, a philanderer, made her pregnant yet refused to discuss divorce with his wife.

Marlene began suffering from insomnia, depression: "It was clear she was nearing breaking point and had lost much of her power to think clearly and control the course of events."[00,356] Whiteman interpreted the situation in terminology to be discussed later; using the house analogy it could be said that the occupant had become a helpless spectator of what was happening there—a powerless participator with an incoming influence set on getting rid of the wife. Whiteman cited the psychiatrist Hermann Lenz as noting that under extreme stress the individual becomes "overwhelmed by some tremendous force from outside himself."[00,361] In Marlene's case it was an uncharacteristic thought of seeing the wife as a removable obstacle to her happiness. She began taking massive doses of Valium (diazepam), at that time not recognised to have highly undesirable side effects in quantity. She evidently entered a dissociated state for much of the tortured time.

As fate would have it, an amputee visited the workshop to say his artificial leg was hurting him. He came from a violent community, and proved quite ready to "do away" the wife for a reward. Months of anguished indecision and failed opportunity to murder passed, with Marlene suffering a miscarriage presumably as a result of excessive Valium and mental turmoil. Her malign inner influence came and went, and in 1974 she arranged to leave Cape Town for Johannesburg. Yet on the day of the murder she found her mind "in a turmoil". An inner "voice" compelled her to visit the accomplice "and see what he feels about it." She drove him to the wife's house ostensibly to tell her the affair with her husband was over. On Whiteman's interpretation of the confused events, the accomplice was sprayed by the victim with a dye-pistol; seized by a fit of frenzied killing he beat her about the head, throttled her and stabbed her with a pair of scissors.[06,373] The case received wide publicity as the "scissors murder".

In a statement the accomplice declared, "The order [to kill] pushed me on. I must now go and kill the woman. I cannot understand myself."[00,371] Evidently malign influences had also entered his "house", resulting in an "insistent directive", which requires that certain actions are performed "even though the individual does not fully understand the purpose, and is also not open to considering their customary or absolute ethical value."[00,113] Perhaps he and Marlene had cross-infected each other with insistent directives by constantly turning over ideas of reward for him and a solution for her.

It takes a special state of detachment to recognise and identify clearly such coming-and-going influences, Whiteman maintained. The culprits were in no state to become detached. Before exploring the nature of such influences, though, events in another murder case that deeply concerned Whiteman reveal more about influences coming and going.

The mother of an adolescent girl, Angela, was killed by Angela and her boy friend, Barry. Angela's plea was: "I deny that any act on my part was carried out voluntarily in that I was acting under the influence of some superior force. And I deny that I was able to act in accordance with my ability to distinguish between right and wrong in that I was unable to resist the superior force referred to above." Barry's plea was: "I deny that any act on my part was carried out voluntarily, in that I was possessed by a demonic force and that I was incapable of resisting the aforesaid demonic force."[00,449]

Angela was born into a troubled marriage which broke up when she was five. She was described by an uncle as "a gentle, kind child ... very considerate ... a very conscientious child, always a very lonely child, but very obedient to her mother." She showed psychic ability; her lonely life (despite a loving relationship with a grandmother) was filled with non-physical companions as playmates.[00,450] Her divorced mother had live-in men friends, one of whom, it was alleged, raped her. She developed persistent severe headaches, blackouts, spells of functional blindness and various other disorders, while psychic abilities continued: precognition, psychokinesis, automatic writing, separative experiences, and the companionship of non-physical entities, although this diminished with time.

Her church-going did not support these abilities, so she sought understanding elsewhere, beginning with the Hare Krishna cult. Eventually she fell under the influence of the School of Wicca and the practice of witchcraft. There is a self-seeking, self-serving side to Wiccan practice which, combined with meeting a particularly malevolent satanist young woman, caused her immense psychological and spiritual disturbance and harm. In a statement after the murder she said felt resentful about her esoteric books being burnt by her mother and her room being "turned upside down". She had been heard by witnesses to talk of killing her mother, yet at the same time she was seen getting on well with her; Whiteman pointed out that no-one talks publicly about murder if they actually intend to commit it.

The boy-friend Barry had witnessed violence at home and was unhappy at boarding schools. He and Angela were classmates. At first she showed no interest in him, but a friendship developed with mild sexual involvement. Barry was concerned about "fits" Angela was presenting among other disturbances, but when he tried to probe the situation all he had were tears, and evidence that she was terrified of the satanist woman, who had taken against her because she refused to be involved with Satanism. This woman threatened harm, and induced thoughts of deadly attacks by "them"; she quite possibly was responsible for psychic attack, apart from physically visiting Angela's mother to tell her Angela intended to kill her. This brought about panic in mother and daughter.

On the morning of the killing, so a statement reads, "it was really an impulse thing to go and do it. That's why we used things like an iron bar and a pan."[00,485] Barry choked the mother by pressing the bar across her throat while Angela hit her over the head with the frying pan.

The judgement in both cases was murder without extenuating circumstances. The concern Whiteman had was the failure—as he saw it—of the courts to understand the underlying mental conditions associated with these two catastrophes. He also saw a failure to analyse the evidence correctly. The trials were not by jury but by a judge and assessors.

Marlene's trial was conducted by a senior prosecutor whose business was to gain a conviction by showing her to be "callous and coldly calculating" in her "planning of the murder". Marlene was not directly cross-examined in court. The Court implicated Marlene directly in the killing, yet Whiteman's analysis of the evidence showed her more as a confused bystander who suffered memory loss after the event. On appeal she was imprisoned for twenty years. In the case of Angela and Barry, both of whom denied carrying out the murder voluntarily, the task of the prosecution was to prove guilt by dismissing the idea of any deep unconscious or paranormal influence.

The first thing one may notice about these cases is that the culprits were all performing out of character. They declared they could not account for their action, that some outside influence had taken control. Swedenborg had much to say about his perception of outside influence. In his *Heaven and Hell* (Ryder, 2010, §292) he wrote,

> Present with every person are good spirits and bad ones... When these spirits come to a person they enter the whole field of that person's memory, and from there the whole field of his or her thought... Several years' continual experience has taught me that this is the nature of spirits' alliance with people. I am surer of this than of anything.

Whiteman saw this "conjunction of spirits" to be a basic structural element of an individual personality; he built this into a wide-ranging concept referred to as "the corporate structure of personality".

CHAPTER THIRTEEN

THE CORPORATE STRUCTURE OF PERSONALITY

The previous chapter featured an idea in the Buddhist Dīgha Nikāya of the personality being as if like a house where people may be seen coming and going. It spoke of "men", evidently meaning that these are definite entities, not formless thoughts passing through the personality. This was confirmed by Swedenborg. Arising from these entities, Whiteman wrote, "we can have thoughts and desires of which we very strongly disapprove. Entire systems of thought and intention, in a strict sense foreign to us, are thus continually 'flowing in' so that we can make a choice according to our unseen core identity."[00,xxiv]

Whiteman attributed the idea of a "core identity" to R.J. Stoller (1973) in the field of sexuality.[00,313] He believed it essential "to recognise that everyone has a 'core identity' in respect of which we say 'I am', 'I perceive', 'I choose', and so on..."[00,xxiv] As a kind of basic component of the house, it is our "lasting disposition", the recognition of having a distinctive being with characteristic tendencies, even if normally we cannot clearly describe its details and structuring. The core identity was seen to be "at first inevitably involved in attachments and identifications (by the third creative function, W). And only when a certain non-attachment has been practised will a holistic release be granted and something of the character of the core identity which is being built up will come through clearly."[06,75]

We may become particularly aware of our core identity when pitted against influences that seem out of character. As with Swedenborg, Whiteman wrote that in a non-physical state these foreign influences "can be *objectified* in a non-physical space." He added, "They may be seen as non-physical living beings, having for us the character of *contributory minds* (co-minds or CMs for short), in relation to which our core identity is an *individual in charge* (IC)."[00,xxiv]

The idea of an "individual in charge" runs against a currently strong body of opinion that denies the existence of a controlling self: "There is in reality nobody in charge, no executive," is how Kelly (2007, p. 640) phrased that idea in rejection of the view. He showed that the idea of self was “absolutely central” to the thinking of William James and Frederic Myers, whose work the book developed in the light of a century of research. To Whiteman the concept of an “individual in charge” was empirically required and logically necessary, the centre of co-mind coming-and-going to form a “corporate structure of personality”[00,xxiv].

As noted in the previous chapter, it requires an advanced spiritual state to experience the corporate structure clearly. While the individual in charge usually fails to be aware of this structure, co-minds also show failure to recognise the structure in which they participate, according to Swedenborg (Ryder, 2010, §292) and Whiteman. “That is to say,” Whiteman wrote, “the IC and CMs are accommodated to each other, without in general realising that this is the state of affairs.” Therefore in order to “Know Thyself”, the IC “must learn to survey component influences in the personality that are usually so closely accommodated to the IC’s situation that they go their own way without being recognised as contributory influences.” They have to be objectified “so as to have knowledge and command of them.”[00,194]

If this sounds far-fetched, it may be noted that an active branch of psychotherapy, Spirit Release Therapy, takes account of this coming and going of entities. It was described by Palmer (2014, p. 1) as having to do with “the possession of an individual by a spirit entity that can be perceived as benevolent or malign, creative or destructive, healthy or unhealthy.” He wrote (p. 1) that Spirit Release Therapy “can be used as a method of intervention to relieve individuals of the influence of unhealthy spirit influences that are often attributed to the earthbound spirits of the deceased.” Complementary to traditional psychiatric models of mental health and religious exorcism, the therapy provides “a dimension of understanding that previous anthropological, theological and medical methods and theories lack.”

While Whiteman was apparently unaware of spirit release therapy, which gained ground only in this century, he cited Freud’s inference of “agencies” influencing behaviour, likewise McDougall, who called them monads, while Balfour, in his study of Mrs Willett’s mediumship, wrote of agencies and co-conscious selves.[86a,27] As noted in the previous chapter, Swedenborg was “surer of this than of anything” that “Present with every person are good spirits and bad ones.” (Ryder, 2010, §292). Myers (1903, vol. 1, p. 16) wrote that our “psychical unity is federative and unstable.” William James wrote of “extra-marginal” consciousness, Jung of “fragmentary personalities.” (Crabtree, 2007). The teaching of a corporate structure of personality can be traced through the Rig Veda, the

Upanishads and dialogues of Gotama Buddha. In the Yoga Sutras (Chapter Twenty-one) they are *nirmāṇa-cittāni,* literally "contributory minds".

Whiteman saw the main thrust of Book 4 of the Yoga Sutras to be the discerning of contributory minds and their clearance towards achieving release. He rendered *sutra* 4 as, "The presence of contributory minds in the personality comes to light simply through an objectifying focus on the sense of self." Then *sutra* 5, "On thus breaking into the onward course of personal consciousness, the one mind of the individual in charge is seen to be the harnesser of the many contributory minds." And *sutra* 6 continues, "In that analysis, any co-mind revealed by contemplation is devoid of attaching effect on the individual in charge." Whiteman's comment was, "The very fact of objectifying a co-mind, or co-mind influence, is to deprive it of attaching effect, however intense the attaching effect on us may have been previously. This is the cardinal principle of release."[93,274]

His translation from the important Kauṣītakī-Brāhmana Upanishad shows "the general state of affairs when CMs [co-minds] are recognised (in a state of release from the physical)."[06,83] Sanskrit terms are not included here.

> To the self oriented to wisdom [i.e. the individual in charge] other selves are brought along and attached, just as to the captain of a team his men are attached. The captain has fruitful experience with the help of his men, and his men with their captain ... He who knows this, having thrown off all wrong [tendencies thereby], attains to self-mastery and higher [spiritual] power.

He believed this to be directly confirmed by his own observations, at least in a separated state. In a letter to me of 10 June 2002 he wrote, "The influence of a CM, and even perhaps a very brief and sketchy view of it, can come to one's attention when the IC is in a 'normal' state. But for clear or prolonged awareness of a CM, one must be dissociated or separated, and thus in some ways different from a normal IC." More strictly, a distinction between an IC and CMs "does not apply when one is settled in a non-physical state. Then there are only 'non-physical beings' with their ruling identities, representative surroundings, and influences guiding or tending to misguide."[03a,106] He termed the observation of co-minds "the objectification of presences", which can occur in any "non-physical state which is temporarily stabilised by some means, voluntarily or involuntarily."[93,33] A full objectification requires entering, in separation, the same non-physical "world" as occupied by the co-mind.

The corporate structure is generally united into a functioning whole in a "normal personality state", but with the right conditions we can focus on what is contributory. "This situation," he wrote, "is like that of an orchestral conductor, able to assess the whole but also able, when necessary, to isolate and assess the parts contributed by individual players."[06,xxiv]

According to this picture, in physical life the individual in charge (IC) is, ideally, the master or mistress of the house of the physical personality, endeavouring to keep true to the sensed inmost disposition of the core identity while trying to deal with dimly perceived intrusive influences. The coming-and-going co-minds are only rarely objectified, and only rarely recognised as being either disruptive or inspirational, as lower or higher CMs.[00,134,546] But the individual, in "habitually identifying with a body and desires that cannot belong permanently,"[06,xxxii] gives rise to "the very general and provisional characterisation that we call the 'personality' of the individual," which "will be said to have then a 'ruling identity'(RI)."[06,82]

Because of identification with CMs, Whiteman noted that a ruling identity can be very different from a core identity.[06,82] In the case of the two murders, the court had the task of passing judgement on the warped ruling identities of the accused, whose core identities had been overwhelmed. And in Whiteman's view the task was carried out with appalling ineptitude. To him this showed a crushing need for a full understanding of core identity and the corporate structure of personality, however complex it might seem.

One may see core identity as a continuing creative substructure behind the "coming-to-be" of an individual life.[86a,183] Being a field of potentiality in relation to the physical world, core identity has a time-ordered aspect, as discussed in Chapter Nine, not fixed but allowing development. The idea of a continuing core identity implies the existence of a line of consciousness, what may be regarded as a time-transcendent, underlying disposition of an individual. Such a line, Whiteman wrote, "can be said to carry all impressions and intentions of which a certain subject, 'I', is conscious—e.g. I am, I see, I think, I can, I decide, I act, I suffer, I enjoy."[03a,105]

While the line of consciousness carrying the core identity develops only gradually, this is not true of the labile ruling identity. It may be regarded as an aspect of volatile individual personality. Whiteman defined "personality" as the "entire organisation of co-minds (CMs) provisionally identified with the IC."[86a,30] Small but recognisable changes in the organisation of CMs occur frequently, and may be said to initiate phases in the personality. Less frequent are major personality disturbances, leading to what is traditionally referred to as multiple personality disorder, although "more perspicuously understood as 'Major Possessive

Disruption' (a CM having 'broken in' and at least partially taken the lead in the personality)."[00,103] These phases may be for good or for ill, such as deviations from true inner core identity that may take the individual to court charged with murder. Whiteman was emphatic that this fluctuating organisation of personality is not to be confused with the underlying core identity or inner disposition of each individual, which develops only gradually.

This gradualness of development, he argued, suggests that a core identity pre-exists any particular physical life and its personality phases. To suggest otherwise "offends against the Principle of Sufficient Reason," since it then "makes an individual come into existence already furnished with a disposition for good or evil, conferred on him by a 'dice-playing God'."[00,36] It follows that "The individual must therefore already have a line of consciousness, with inner memory and corresponding disposition worked out when he/she takes on the new physical life."[00,139] The idea of physical death being the end of the line "seems to rest largely on ignorance or rejection of the evidence for 'other worlds' and of processes of separation from the physical body, as well as any logical purpose in regard to the continuing life of individuals."[06,75]

Whiteman considered that after physical death the term "individual in charge" no longer applies to a former IC with its memories of the former personality. The associated CMs that have become detached can also carry these same memories. These CMs, as well as the former IC—now with the status of a CM—"can temporarily enter the personality of a still-living IC, if there is sufficient affinity and need."[00,139] They function as temporary co-minds in the new personality; on the other hand a former IC may enter a new physical life permanently as its IC, effectively continuing a line of consciousness through the physical lives. These complexities have to be taken into account in interpreting what is generally taken to indicate reincarnation.

CHAPTER FOURTEEN

REINCARNATION AND ALTERNATIVES

The previous chapter noted the possible intrusion of one or more memory-holding CMs into the corporate structure of a physically-living individual. This tends to be interpreted as evidence of reincarnation if the carried memories are thought to be intrinsic past-life memories of the new physical person. Whiteman treated the subject of reincarnation with great circumspection. Considering the compound, shifting structure of a whole physical personality ("a profusion of loosely linked tendencies"[86a,121]), he maintained that "it is wrong then to say that we, as personalities in the world, might be reincarnated."[06,x] This cannot literally be the case, he argued: "For any personality consists of an IC together with CMs, and the former develops, while the latter change, continually. Hence the personality in the present scene cannot be the same as a personality in a scene of the past."[86a,124]

Yet if a former IC returns to the physical world as the individual in charge of a new personality, continuing a line of consciousness, Whiteman was prepared to apply the term "strict reincarnation". The term "loose reincarnation" was then applied to the temporary intrusion of a CM into a personality structure. Whiteman saw loose reincarnation largely to account for the reported cases of a young child presenting apparent memories of the life of someone who had recently died. Because the CM is not the individual in charge, "As the child gets older, naturally that CM has to retire, so the child loses those 'memories' entirely."[06,xxxvii]

He cited passages in ancient Indian literature that show awareness of the difference between strict and loose incarnation, for example, a passage from the Kaṭha Upanishad: "some spiritually embodied beings enter a womb for physical embodiment [i.e. strict incarnation]. Others assemble successively on a 'stock' according to their deed-attachment (*karma*), according to inner instruction [i.e. coming-and-going loose incarnation]."[93,32]

Whiteman emphasised that any intrusion of memory items and of disposition into a new life, suggestive of reincarnation, could have several different sources garnered from fragmented existences. To begin with there is his distinction between loose and strict reincarnation. His complex view may be compared with the more standard view of Stevenson. In what appears to be missing correspondence, Whiteman evidently raised the idea that physical personality is composite, and that on dissolution after physical death only a fraction of one personality may carry memories into the composition of a new physical personality. Stevenson replied in a letter dated 17/4/1963, "so far as the empirical evidence available to me goes, the evidence seems rather to support the beads on a string idea of the theosophists." He evidently saw an earthly personality to be a discrete unit, connected by a discrete link between one incarnation and the next. Stevenson's more usual beads-on-a string concept was linear; in contrast, Whiteman's conception was reticular, a network of entities connecting lives.

In a letter to Stevenson of 17/4/1987 Whiteman wrote: "According to Swedenborg and the Buddha (and in my own experience) a personality *does not exist* without co-minds ceaselessly coming and going ('no man is an island')." This issue weighed with him because of his own observation in separated states of a corporate structure of personality. In contrast, Stevenson's approach was suppositional and theory-based, adopting a "beads on a string" idea said to come from Theosophy, a movement from which Whiteman had withdrawn. The "string of beads" idea must have seemed to Whiteman out of phase not only with the original teaching of Gotama Buddha but also with the early Upanishads (on which Theosophy is supposedly based).

Stevenson did admit (17/4/1963) "that there are few cases which are not also open to the other idea of a partial incarnation of members of a group. The cases which, it seems to me, can only be explained by strict reincarnation are those in which birthmarks occur... In these birthmark and deformity cases, the link between one personality and another seems quite clear, assuming authenticity for the cases." Whiteman's reply to this is full of the complexities that he saw in the whole matter of personality structure and lines of consciousness. He believed that the occurrence of birthmarks did not rule out the possibility of loose reincarnation, the influence of a co-mind being potentially as powerful in an infant as is the core individuality. He wrote (17/4/1987), "if one accepts that co-minds (still attached to experiences in the life they have recently led) will necessarily help to make up the personality of a baby just born (in fact may constitute it almost entirely if the individual in charge is not fully awake, so to speak), then naturally one would expect traumatic events in the life of those co-minds to be manifested in the physical body of the baby."

The existing correspondence sadly does not contain a debate on the issue. It seems that Stevenson himself remained unconvinced by this multiple "picture of the personality" and all it entailed. At the end of his final magnum opus (1997) he seemed still to have adhered to a "string of beads" idea by proposing a theoretical entity, the psychophore, "a vehicle for memories between lives." It seems that Stevenson found Whiteman's account of the complexities of personality structure and its vicissitudes after physical death to be unnecessary, not having had direct experience of such things; he must have felt he had enough complexity to deal with as it was. On the other hand Whiteman believed it was of cardinal importance to understand the complexities of personality structure, being part of his direct experience.

This difference of opinion underlines an important fact: the extent of personal experience governs almost completely one's view of what counts as legitimate and useful knowledge. Where direct personal knowledge is lacking, the device normally used in hypothetical-deductive science (Chapter Eight) is of setting up theoretical entities such as a psychophore. These may help to focus thinking, yet they tend to create an obscuring "cloak of ideas", as Husserl put it in his phenomenology (Chapter Eight).

By early 1987, discussion between Stephenson and Whiteman on reincarnation became involved with what Whiteman saw as "one of the greatest difficulties in this field of research", which was "distinguishing between the entering into a retrocognitive scene, and the plausible identifying of details in the scene with memories in the same 'line of consciousness as the presently-living percipient'." (12/1/1987) In other words how does one distinguish between an apparent memory taken over retrocognitively from an unrelated past, and an intrinsic personal memory suggesting reincarnation? Whiteman claimed to have had several such "tapped in" memories, as will be discussed in the next chapter. Stevenson replied (7/4/1987) that he had "little of value" to say about this distinction; it could "account for occasional cases of the type I have studied, but certainly not all of them; it does not adequately explain the occurrence of birthmarks and birth defects corresponding to specific wounds in a deceased person."

Whiteman later concluded that these difficulties of interpretation regarding apparent childhood memories "do not amount to a denial of reincarnation—only that reincarnation cannot be *proved* that way."[06,113] He nevertheless believed it to be possible for there to occur "a series of credibly *objective* episodes in which we are apparently the same person." He cited a series of experiences in 1974:

> I was an apprentice Minoan priestess, as I put it in my own language, and puzzled objectively about certain particulars, such as the absence of petticoats beneath the conical skirt, and the Minoan understanding of the

> "sacred marriage". About three years after this I discovered, in a book not published at the time, a picture from which one can see that the conical form of the skirts worn by the Great Mother and her attendants was secured by a heavy circular wire at the hem, and no petticoats were in fact worn.[86a,123]

While Wordsworth wrote of trailing clouds of glory into a new life, evidently darker clouds may be carried too. This raises questions of responsibility and culpability in the life of an individual, which will occupy much of the following chapters, with Chapter Twenty-one considering the specific issue of *karma*. This present chapter may be ended by noting that Whiteman was careful not to attribute all apparent past-life recall events to a form of reincarnation, either strict or loose. Many complexities arise in thinking about lines of consciousness, which require another chapter.

CHAPTER FIFTEEN

TANGLED LINES OF COUNCIOUSNESS

Whiteman discussed a number of possibilities in which "possession of a memory by no means indicates that one personally experienced, in the physical world, an original of the events 'remembered'."[86a,124] Some presentations of past scenes can, he maintained, be better explained not by reincarnation but by a form of retrocognition, which he termed "mediate identification of the past".[86a,120] The term "mediate" was intended to imply "the utilising of a personal life and memory quite different from the life and memory of one's physical personality, for the purposes of self-fulfilment through a new and instructive identification." The term "identification" was used to mean "the setting up or taking on (as one's own) of personal characteristics of some kind." (Letter to me of 15/11/2003.) This may be in physical or non-physical life.

He found that "The sense of really living those events can seem sometimes far more intense, more sensitively responded to, and better understood, than are similar events in the physical life."[86a,116] Yet because these events occur in secondary separation (Chapter Four), "one has also a 'background' faculty which assesses the experience, tries to relate it to one's physical store of knowledge, and usually attaches words in order to establish a secure memory of the details."[86a,117]

He gave several cases where, in a state of secondary separation, "the scene entered by the percipient is, to all appearances, in the past, though acted out in the present." The following experience of his was recorded in 1971.[86a,121]

> While lying in bed I entered suddenly a state of secondary separation and found myself standing upright, wearing a long dress, with my feet in shoes just showing below, and also a conical headdress of the kind called a henin which was fashionable in England from 1430 to 1485. One part of me was vividly aware of the feeling of the headdress, perfectly natural and indeed familiar, while another part, in background thinking, was surprised that such a headdress should be comfortable and satisfying to wear.
>
> At the time, I firmly associated this kind of headdress with 12th or 13th century England, and called a "Plantagenet" headdress. My "background

> thinking" tried to impose this interpretation, but it was as if a voice said, peremptorily, "No, Italy!" With this there came a strange but definite "flavour" of medieval Italy, as known by myself "here and now", while the background thinking of myself living today knew it as "in the past" and placed it in the pattern of historical events. The name Florence was also put into my mind.
>
> Later, in an endeavour to find the date of the fashion and whether it was also taken up in Italy, I searched in the university library for evidence. I eventually discovered, besides the facts mentioned above, the reproduction of a plaque showing a "Giardino d'Amore" with three young ladies all wearing the henin and marked "Incisione Florentina (1460)".

He emphasised that the clarity of this kind of experience might induce many people automatically to assume that the scene of the happenings "reproduces a true memory of their own in a previous life, because it is 'of the past'."[86a,122] But he pointed out that this is not necessarily so. In the case of a good object-reader (or psychometrist) the vivid entering into a past scene cannot be considered entry into a previous incarnation; the number and variety of instances makes this idea "preposterous". Another difficulty of a reincarnation interpretation arises from a "switch over" during secondary separation, from being one person in the scene to being another. He cited instances in Joan Grant's *Far Memory* and Rachel Levy's *The Phoenix' Nest* where, in claimed memories of past lives, "the character was first seen most vividly from outside and then immediately known from within." His comment: "Obviously one cannot have been incarnated in two different people living at the same time; and preference for one of the characters does not establish a historical identity with him or her."[86a,123] So "if one retrocognitive 'entering' cannot be a genuine memory, then obviously others might be similarly not genuine [of reincarnation]."[06,113] Either or both recalls could have been contributed by temporarily attaching CMs.

It seems that in such cases there is some intertwining of life patterns, which makes a simple reincarnation "story" farfetched; this places other accounts under scrutiny. Whiteman seemed to be undecided about the interpretation of his henin experience. Yet, in separations scoring high on the Reality Index (Chapter Four), he nevertheless believed it to be possible for there to occur "a series of credibly *objective* episodes in which we are apparently the same person." He cited the Minoan series of experiences recorded in the previous chapter. The sacred setting of the experiences might have contributed to their high reality rating, apparently of strict reincarnation.

In addition to mediate identification of the past, Whiteman also recognised a tangle which he termed "mediate identification of the present". This could result in entering the physical life of a living person as a CM, or, he believed more likely, entering the life of someone recently deceased, not physically alive, again as a CM.[86a,119] Here is a short example he gave in secondary separation of an apparently presently-happening event:

> Seated in a café at a round table, with a young man. He questions me about another girl, named Jackie, hoping that I can help him to understand her behaviour. He says, "It's impossible to understand what she wants." I return to physical wakefulness for perhaps 20 seconds, thinking over the experience and the character of Jackie as I "remember" her, and then the same scene opens out again, the conversation not having been interrupted. He says, "Do you think you could do something with her?"[86a,117;00,202]

He found the character of Jackie "extraordinarily clear", but we are not told if anything was resolved. The point to be taken is that the coming and going through the house of personality can take many paths, some not easy to follow. He cited the important fourth "over-knowledge" (*abhiññā*) of Buddhist teaching, "in which a series of past lives (not necessarily one's own in any way) is briefly entered, and one acts out certain scenes or otherwise makes observations there. More commonly, only one retrocognitive scene is carried out. Reincarnation is not implied."[00,87]

Mediate identification can be involved in the phenomena of apparitions, discussed in Chapter Three. In mediate identification one can, in separation, "take on the bodily appearance and relevant memories of another living being;" but conversely, "a CM normally with us can take on our bodily appearance, many of our memories, and even a strong intention that has taken hold of us; and it may then act out that intention as an apparition to someone in the physical or some 'inner' world."[03a,108] He termed such a CM a "double", "that is, CMs who have taken on the physical appearance and many of the memories of the individual who is 'doubled'."[06,27] Citing cases from Gurney *et al.* (1886) and Tyrrell (1953), he noted complexities such as, "There are in fact a large number of cases where an agent has succeeded in sending a double of himself as an apparition to someone else at a distance. The agent may or may not have some impression of being at the place where and when the apparition is seen, but is hardly ever aware of what his apparition is doing."[03a,109]

As Whiteman perceived it, this situation is highly complex. While the apparition may be regarded as a double, strictly speaking it need no longer be regarded as a CM, "since the non-physical being in question has escaped from the personality of the individual normally 'in charge'. But it may seem to return, and in that case it is convenient to speak of it still as a

CM, even when separate."[06,27] The complexity increases when we need "to take thought when a human figure resembling someone in this world, or dead, appears in a separated or dissociated state. For what is seen could possibly be a double, and thus a *former* contributory mind belonging to the personality of the individual in question, but now no longer so."[03,28] The double may be mimicking a former individual in charge, and so give a false impression of the IC's survival.

The complexity that Whiteman presents has its origin in the concepts of a corporate structure of personality, with contributory minds "coming and going", also with CMs that form doubles appearing as apparitions. All this is combined with his view that apparitions manifest in a strictly non-physical sphere that requires a state of dissociation or separation to witness them. This was discussed in Chapter Three; his view was that openings, which include seeing apparitions, occur in a spatial field that is not physical. Whether Whiteman provided a workable handle on this complexity could be questioned; what cannot be questioned is the inadequacy of a one-level naturalism to account for it.

Another tangle of lines of consciousness, one that directly concerns the two murder cases, was treated by Whiteman as "psychic infection". He wrote, "Many kinds of experience confirm that powerfully motivated influences can act *as if from a distance* on the mind or body of a physically-living individual, with or without his/her willing co-operation."[06,429] As the influence is non-physical, physical distance is not a factor. Influence can be benign, as in psychic healing, neutral as in channelling and mediumship, or malign.

Malign influences may be suspected especially when an individual acts out of character in a harmful and criminal way. Infection with an "insistent directive" is notable in the case of Marlene's accomplice. Whiteman observed that infection may not be simply a mental state and intention that is deliberately imposed by an agent, presumably Marlene in this case; "the individual from whom the directive seems to originate may be quite ignorant of its nature, such individual being in a dreamlike depersonalised state and him/herself subject to some deeper influence."[06,430] Further, the directive comes and goes fluctuatingly in each individual, due, Whiteman believed, to the possessing CMs coming and going. When the influence is not active then life continues normally; the individual tends to forget it as if it were a dream. This adds another degree of complexity to evaluating responsibility and guilt.

Nevertheless, Whiteman considered that all power of choice is not generally lost in a fluctuating state of infection or possession. Even in a state of mental black-out, "the subconscious, elements in one's personality

(CMs) may object to what is going on and cause a resumption of consciousness. So even in such a case, the confirmed disposition has an effect not very different from the workings of conscience."[06,442] This implies that contests are continually taking place within the individual. Indeed, all of what has been discussed about the coming and going of influences and their impact on an individual, highlighted in the two murder cases, points to a central idea in Whiteman's work: the inner contest.

CHAPTER SIXTEEN

INNER CONTEST AND SKILLS

Whiteman's trilogy, *Old and New Evidence on the Meaning of Life*[86a,00,06] is subtitled *the mystical world-view and inner contest.* "The personality," he wrote, "is the field of contest, while the individual stands in the midst, subject to both [wise and unwise] trends at once, in an indescribable fusion."[86a,211] The contests were seen to be infinitely varied and continual in life, conflicts between "the unwise impulsions of self and inklings of wisdom,"[00,117] of right-doing and wrong-doing, or between resolve and weakness, or between Freud's Reality Principle and Pleasure Principle. Contest involves stress, which involves "the refractory force in general taking hold of a person, and either faced with faith and so brought to a transcending level, or escaped, voluntarily or involuntarily."[00,6] Yet stress is "the only way in which our development can be initiated and secured."[06,63]

Whiteman's fullest treatment of inner contests is in Chapter 20 of Wolman & Ullman's *Handbook of States of Consciousness* (1986), which was reproduced as Chapter 8 in Volume 2 of *Old and New Evidence on the Meaning of Life* (2000). Possibly easier of access, *Handbook* references are included as [86b] together with the 2000 references.

In contests, "outward tendencies to attachment, fixation, discomfort and awkwardness, automaticity, or anxiety-provoking stress are countered by 'higher' motivations that are secretly working for the development in us of corresponding powers of release and control."[86b,630;00,159] This development takes place through the cultivation of skills, which need to be discussed before considering contests where the skills come into play.

Much if not most of the cultivation of skills is through learning processes during contests between what proves to be helpful and wise, and what is not. The skills ultimately lead to control of the four creative functions (Chapter Ten) which operate in life, and so to a release from debilitating attachment to self, things and ideas. Attainment of non-physical and mystical states is not possible, Whiteman believed, without the cultivation of the appropriate skills. Essentially these are skills of reflective relaxation, watchfulness, and the ability to objectify feelings, thoughts and incentives to the point of saying, "I am not this, this is not mine, it is merely something which flows into me, and I observe it coming, present, and then going, all without attachment."[86a,17] This is a formula for successful psi performance as well as for more exalted experience.

Whiteman made an important distinction between two chief degrees concerning the "ability to objectify". This comes particularly to the fore in considering culpability in the two murder cases. In the first degree, feelings, thoughts or incentives are perceived as if on a TV screen, and we can identify ourselves with them or not as we please. This was classified as "a higher or releasing *dissociation* or it could be simply called a *non-attachment*." In the second degree of objectification, typical of secondary separation (Chapter Four), "we are released from the thoughts, feelings and incentives in a far more complete manner—as if we were actually in the TV studio and saw the persons responsible for these expressions of personality, located apart from us with indubitable reality." This skill of freely watching within ourselves, Whiteman maintained, is regarded in Buddhism as "penultimate in the mystical way," and "involves some kind of 'collectedness' within the personality."[86a,18] Exercise of the skill should have diverted the fateful course towards the two murders; it leads to considering the practice of Recollection, introduced in Chapter Nine.

Active and Continuous Recollection were regarded as being two of three "foundational skills." The third was termed "Faith (happy confidence in *Truth*) and Obedience (untroubled response to *Good*)."[93,247] The skills were linked to the four creative functions from east to north: Faith (E), Obedience (S), Application (Active Recollection) (W) and Non-Attachment (Continuous Recollection) (N).[86a,22] The term Obedience is easily misunderstood. It covers a range of levels: "For example, mystical Obedience (to the archetypal Right and Good) has an initiatory form called 'duty', and a neutral form called 'compliance'. 'Servility' is a counterfeit; and 'self-will' is an opposite, but can also be seen as a corrupt Obedience, turned to compliance with urgings arising from self." [00,4] Wisdom becomes attached to obedience as "an observing of the higher values of wisdom and a willingness to act according to them."[00,48]

This observance is a well-recognised ingredient of mystical experience, noted by William James (1902, p. 293) when "the mystic feels as if his own will were in abeyance, and indeed sometimes as if he were grasped and held by a superior power." Yet, because of the importance Whiteman attached to the term obedience, a more definite statement is needed. He recognised three degrees of obedience:

> The first is reliance on formulas of moral law or social custom. The second

> is what may be described as a "normal conscience" ... In the third or mystical degree of Obedience one must first ensure that one has an open mind, well-informed upon all facts that appear relevant, poised and free from conditioned beliefs or desires (e.g. wishful thinking, partisanship, or other unsettling prejudice). Then, when one is at last settled in a state which does not depart from orientation to the right and good (as shown by equanimity and other signs), one waits for an intimation that some contemplated action to fulfil the right and good does in fact conform with what is right and good. And no action is consented to, nor is inaction consented to, unless a well-grounded orientation to the right and good (or as we may say, to Wisdom) is maintained in doing so.[00,29]

The disposition towards Obedience was nevertheless recognised to differ in the two sexes. Discussing "core sexual identity" (Chapter Seventeen), he noted that in the female the essence of Obedience "must be declared perpetual, at the root of every action." The male is characterised "chiefly through the disposition to Obedience being auxiliary and occasional."[00,207]

By Faith was meant "an inspirational enlightenment and energy which ... springs from what is wholly unconditioned." Yet it requires a "continually watchful orientation and adaptation to an ultimate Wisdom," which is "the function of a true selfless Obedience, also grounded in what is wholly unconditioned, and also ultimate in character."[86a,20]

One may recognise in Faith the actions of the first and third creative functions, and in Obedience the second and fourth functions. So, "Faith (first function) looks outward to action (third function), while Obedience (second function) looks outward to security (fourth function), and each serves to the other as its 'auxiliary'."[86a,211] The first function may be seen to involve obedient Faith, the second, faithful Obedience. "For Obedience is 'of faith'—it must have faith in a universal Wisdom to which it responds with obedience. And Faith is 'by obedience'; for without a secret origination for obedience to Wisdom it would not be a true faith."[00,51] As Whiteman pointed out, Faith may be characterised as masculine, and Obedience as feminine. This is in keeping with the general character of the first and third functions as masculine in contrast to the second and fourth feminine functions.

Whiteman considered Obedience to emerge from habitual Continuous Recollection, orientated to Right and Good, Wisdom and Reality. He believed that "An orientation of Faith cannot, it seems, be practised directly, but must be 'born' from Obedience."[86a,11] Therefore Obedience is more prominent in his work than Faith. Nevertheless, neither is passively implanted in the personality but has to be developed through the stress and effort of inner contests. This involves living through lows as well as highs of experience: "one cannot know states of low development in an objective and wise way (so as to be safe from lapsing into them) without first experiencing them in a subjective way. We must *rise,* through contests, along a *scale of upliftment* from unwisdom to the fulfilment of wisdom."[00,30]

Contests are between opposites. The opposites need not only be polar, like cruelty/kindness or hate/love, but may involve what Whiteman termed "counterfeits". He wrote, "Thus instead of perseverance we may have obstinacy; instead of wise and selfless response, we may have gullibility; and instead of patient over-ruling and transcendence of frustration, we may have an impatient and unbalanced over-riding of it."[86a,210] The conquest of counterfeiting is attained by the three foundational skills, also called voluntary powers,[00,51] so "Active Recollection replaces automaticity and indulgence by wise control over action and sensing; Continuous Recollection replaces conditioning and complacency by true security and contentment; Obedience and Faith replace self-will and its justification by selfless wisdom from the Divine Source."[86a,211]

"Counterfeits and opposites," Whiteman wrote, "can be identified as the refractory elements in any contest, responsible on the one hand for 'bad intentions' in any personality of this world, and on the other hand for all the strange and disrupting manifestations of psychopathology."[00,54] Among refractory elements, Whiteman mentioned the medieval Christian triad—the world, the flesh, and the devil—and a similar triad in Buddhism: *bhava* (becoming, the world), *kāma* (sense-pleasure) and *avijjyā* (unwisdom). He also quoted a similar saying by Eckhart: "Three things prevent a man from knowing God [the supreme Source and Universality of higher values] in any way. The first is time, the second corporality, the third multiplicity." He interpreted this as, "'time' suggests compulsive flow, without time-stopping focus on any detail; 'corporality' suggests an unreflecting wholeness; and 'multiplicity' suggests disunity of motivation." These are overcome by the voluntary powers of Active Recollection, Continuous Recollection, and Obedience respectively.[00,54]

The "unreflecting wholeness" of corporality was seen as part of the fault of being subject, on the one hand, "to the fragmenting effect of details too exclusively grasped at, and on the other hand to wholes that are only hazily glimpsed at and so open to ambiguities and misconception." The particularising "left" and the holistic "right" must therefore be synthesised, "gained through contests with the effects of fixation or rough and ready opinion. Such contests can be said to shift us along a scale of right value-building."[00,31]

Whiteman saw co-minds (Chapter Thirteen) to be involved in these contests as higher or lower "influencing entities". He wrote that although these entities "may exceptionally be 'seen' non-physically, they should evidently be thought of as primarily *sources of influence,* communicating beliefs, intentions, feelings, and sometimes sensory impressions loaded with mental persuasion."[00,59] Lower CMs impose counterfeits and harmful impulses, higher CMs fight for what is "genuine and free."[86a,211] The house of personality is therefore not simply a thoroughfare for influences but also a battlefield. This is nowhere more devastating than in cases such as the two murders, to which we may return after considering stress in more detail.

In the course of spiritual development, nearly all contests involve some measure of stress. They may be faced, or shelved by digression, or escaped either by some form of personality disorganisation or by deliberate choice of avoidance.[86b,631;00,160] Yet eventually they should be engaged and lead to aware spiritual development: "all skills and spiritual powers which make up a controlled and unified character of the individual are developed in a discipline of 'warfare' with refractory or opposite conditions."[00,50]

In the case of lesser contests, "the development of Active Recollection as against automaticity, and distinguished by disorganisation, or of Continuous Recollection as against conditioning, and characterized by neurotic anxiety ... have no ultimate or end-of-the-world quality."[86b,645;00,175] But in contests associated with "disclosure of the Divine Source and mystical transformation",[68a,646;00,175] trials and upheavals may have an "end of the world" quality, since transcendence means elimination of lower CMs and resulting destruction of the old personality. This "end of days" destruction can present outwardly deranged episodes; "violence or other egodystonic behaviour may be acted out. At other times there may be gratifying uplift or steady excitement communicated by the glimpses of strange spiritual powers and release from the artificialities of physical life."[00,58]

Where such disruptive episodes are "objectively watched and evaluated, without fear, and in full understanding of the spiritual character of what one is being taught," Whiteman applied the term "ultimate contest ruled by Obedience."[00,58] It was termed ultimate because "there is no 'escape' from such contest into a deeper dissociation, and there are no higher functions (than those primarily engaged in such contest) to cultivate so as to gain control."[85,30]

On the other hand, in states "in which little is understood and unpleasant things cannot be shaken off," the term applied was "near-ultimate contest."[00,58] In either case there is restructuring of the personality: "It is clear that the ordinary 'person' who has grown up anchored in the world, has been partly or wholly done away with, or transcended, and a new and *real being* is either in charge (through Obedience) or is hoping and trying to get known, against forces aimed at destroying it."[86b,651;00,182]

These upheavals, which may include shamanistic phases, were perceived by Whiteman to appear in allegorical or mythical form in many passages of the Old and New Testaments, as "last times" or as an apocalypse. These are discussed in Chapters Twenty-four, Twenty-five and Twenty-nine.

In normal development, the object is to control a contest, "to replace stress by ease, without giving up one's intention."[00,43] Release comes by "letting go", but this may be countered by self-created attachments caused by "fixed ideas, formula thinking, uncontrollable habits, inability to act holistically, or an emotionally grasping attitude to whatever backs up the fixation or seems to relieve the stress when the fixation begins to be exposed."[00,45] Control of the contest then fails, and the individual is trapped in a cycle of recurring failure. About this failure, Whiteman commented, "although the influence itself (in favour of the fixated cycle) can be called 'evil', the motive in one who is so possessed is 'mistaken' rather than 'evil.'"[00,50] This was a point not taken in the trial of Marlene and Angela.

The issue is different in cases of psychopathology, where control of refractory elements is overwhelmed by disorganisation, neurotic escape or psychotic escape. These are kinds of "involuntary escape" in a dissociated state, which can give rise to symptoms of mental disorder. In psychotic escape, Whiteman believed that "lower entities" or CMs assume control of the personality, while "the individual is largely recessive, in some kind of possession state at least most of the time, and perhaps sometimes taken into a black-out."[00,57]

Even so, possessing entities "can be revealed to more advanced beings, and thus to some extent to the individual with whom they are fused. In this way the individual may watch as in a dream, without control or sense of responsibility—the state is depersonalised."[00,57] Further, there may be derealisation, in which the surroundings seem strangely unreal and possibly distorted, while conscious awareness fluctuates into occasional haziness or black-outs. These dissociation states he believed to be signs of possession, which one may recognise in the cases of Marlene, Angela and Barry.

With possession, inner contests can be said to take a more complex form. Whiteman recognised "possession" as "The direct perception or well-supported inference of a controlling power on the personality or physical body of a certain individual in charge, such power being exercised by a meaningfully intentional force or non-physical living being." The controlling power "conspicuously lacks normal compatibility with the IC's core identity, and cannot be easily brought into harmony with it." [00,130] He recognised thirteen different possession states, varying from a condition of benign channelling or mediumship through to malign Satanism.[00,106]

The lack of an IC's control in these states is likely to result in dissociation, if not some degree of depersonalisation, and possibly derealisation.[00,64] This was in evidence in the murder trials—time and space were in a mess, as Barry said. Yet on the positive side, Whiteman held that it is only in a state of dissociation or separation that an individual, through consociation with higher CMs, will attain freedom from stress and a stage of release "in which ties and resemblances to the physical body and the normal physical life are put off, though still available to the memory."[00.64] This is the beginning of what he termed mystical transformation.

If a dissociated state is a point of departure both for symptoms of mental disorder and for mystical transformation, then clear analysis is needed. The former happens when contests are not successful and the individual is trapped in a cycle of lower order, self-engendered fixation. The individual becomes subject to compulsions of a counterfeit moral character and insistent directives. Whiteman pointed out that such a state of affairs "occurs characteristically in adolescent breakdown, and is in general easily recognised from the bizarre character of the phenomena produced by the possessing lower CMs, which cannot be shaken off."[00,71] The murder cases were seen as examples of this.

On the other hand, "When the IC has developed the power to remain in a dissociated or separative state of higher order, he/she will obtain spiritual body expression through consociation with higher CMs with whom there is complete harmony." Then, at a certain stage of release, there occurs the beginning of mystical transformation. But, Whiteman noted, "that can happen only if an orientation of Obedience has become habitual."[00,64]

The expression of a lower state differs in males and females, along with many other actions and responses within the house of personality. Whiteman believed the ways of dealing with intolerable stress are "almost always pursued uncontrollably by males and females (3rd function disorganisation for males; 4th function dissociation for females).[00,532] Whiteman's investigation of these features needs to be considered when taking account of the two murders, and of spiritual development in general.

CHAPTER SEVENTEEN

SEXUALITY AND CHAKRAS

Inner contests and sexuality are inextricably mixed, Whiteman maintained. But his was not "a biological view of sexuality in which labels ('male' or 'female') are attached to certain structures of the physical body and then, derivatively, to the person, without reference to any *experienced sense* of maleness or femaleness."[00,215] Instead, he recognised four different levels of experienced sexuality: 1) biological structure and function, 2) normal accultured sex-identity, 3) obsessive fixated desires or habits, and 4) intelligible sex (core sex of the personality, ultimate sex disclosed spiritually). [00,195] The four different levels are not necessarily all male or all female. "An individual who is male (or female, as the case may be) at any one of these levels may appear at another level, temporarily or permanently, as female (or male, in the other case)."[00,570]

This idea of potential bisexuality is needed to understand spiritual development, he believed. Central to the concept was that "Neither sex can develop spiritually without any form of participation in the life of the other sex... Bisexual effects in the manifested personality can be easily explained, following on evidence for the corporate structure of personality [Chapter Thirteen], by accepting that contributory minds (CMs) predominantly of one sex come to the fore in the personality on occasion, and may be male or female. Normally these predominant CMs will be of the individual's biological sex, but not necessarily so."[00,193] Therefore "an individual who is normally male physically can be female in core identity, and vice versa, in each case with a normally balanced personality (secured by CMs of both sexes), everyone being thus bisexual in full personality."[06,83]

Whiteman's view was based on his own life experience. He wrote:

> Much preparatory discipline with recollective practice and otherworld disclosures had been given me during 24 years prior to ultimate disclosure of the Divine Source [the "Right and Good"]. Within a few seconds then I was acutely aware of being, as it were, a girl of about 14 years old, wholly and permanently female, in an intelligible, ultimate, and lasting way, beyond all shadow of doubt. This perception came almost entirely from

> awareness of an ultimate core-disposition, which I forthwith named Obedience. That meant a loving, trustful, and virtually permanent orientation to the Right and Good, characterising that ultimate femaleness.[00,222]

Mystical transformation of this kind commonly includes the experience of being an age between about seven and fourteen,[06,194] or seventeen in his case.[61,231] For someone having a female core identity, appearance as a girl would be normal in separation. Any physically-based psychological notion that this indicates a transgendered individual runs into a radical category mistake, conflating the physical with the non-physical. Using Husserl's term, such physicalistic notions have to be "bracketed off" as irrelevant and inapplicable. Whiteman repeatedly emphasised that an appearance in separation as core identity in no way resembles the physical appearance of a physical personality. In his case, physically there was no sign of femininity in his behaviour or undertakings, although he recorded, "As a result of disclosure of the Divine Source, and its transforming power, I am very often carried into some degree of tertiary separation [Chapter Four]; that is to say, I am aware, dimly or vividly, of characteristics of my spiritual-bodily form at the time." He added, "it was indubitably clear that my 'real being' was created archetypally female."[00,320] Further, "It was not that I (in some other nature) observed a femaleness, but simply that I and femaleness were revealed as the same thing, while maleness was something that I could think of and even feel, but which did not belong in the same absolute and perpetual way."[00,321]

Yet even in ordinary physical life, Whiteman wrote, ability of an individual to think and feel in the mode of the opposite sex results from the real presence of the potentiality of the other sex in the individual. He termed this the "auxiliary sex". An instance of witnessing his auxiliary sex in tertiary separation was described when, four months after the feminine disclosure, he had the experience "in which I was identified with 'the Son' for about ten minutes."[00,322] But this did not have the ultimate and lasting character of disclosure of being "archetypally female". In reviewing mystical writers in whom "the Son" (archetypal male) appeared ultimate, he listed Isaiah, St Paul, Eckhart, Swedenborg. The archetypal female "daughter, bride, mother" appeared ultimate in St John of the Cross, St Bernard, Chaitanya.

Whiteman noted a recognition of "bisexuality of motivations" in the researches of Freud and Jung,[86a,20] but he did not equate his conception of the female auxiliary sex in men with Jung's conception of the anima. He cited Jung's description of it (1963, p. 349) as "unconscious, dark, equivocal and purposeless."[86a,108] While acknowledging that the female auxiliary sex is rarely consciously perceived and admitted in men, Whiteman saw it as being potentially enhancing, not "the deplorable condition of 'the woman within.'" (Jung, 1964, p. 31). Evidently Whiteman was correct in judging that "behind the phenomena lay a psychological situation which Jung was not able to face in its true objectivity and therefore could only be introduced to in a symbolic form."[86,108] This seems to be the general condition in men.

In Whiteman's view, "for spiritual health the 'feminine' functions (2nd and 4) must be balanced with the 'masculine' ones (1st and 3rd), none being counterfeited or undeveloped."[86a,108] The presence of both principal and auxiliary sex in the individual is thus reflected in the four creative functions (Chapter Ten) which can contribute to a fully integrated personality. The first and third functions—purposive energy, action—are masculine principal functions; the second and fourth functions—feeling, evaluation and choice, expansion into selflessness and security—are feminine principal functions. Thus the auxiliary masculine functions are 2 and 4; 1 and 3 are auxiliary feminine functions[00,290].

Ideally all functions would be operating in conscious harmony and effectiveness. An individual would then be comfortable in having his/her auxiliary sex empowered. He wrote that "the functions of one's auxiliary sex can be knowingly utilised, as accessory functions, or in 'acted out' identity at the physical and other levels. The bisexuality of the creative cycle requires that life in full awareness of one's principal sex could not proceed without a cultivated ease in utilising the functions of one's auxiliary sex in an unmanifested or provisional way."[00,222] Even to the extent of "acting out" in cross dressing, "the auxiliary sex will have grown sufficiently familiar for the individual to rest happily in it, at least for a period of time, through social adaptation; and the stress which originally led to the transvestism (as an attempted escape) has been thus largely overcome."[00,258]

In girls, special perils are liable to develop during adolescence, as shown in the case of Marlene. Whiteman examined this complexity through an analysis of the corporate structure of personality and inner contests. He provided evidence from a number of sources that in the eight to ten years before adolescence, development of a girl's auxiliary male functions tend to be dominant; adolescence is the time when her "essentially female functions must come more fully into play and be developed accordingly."[00,261] He cited Mary Pipher's findings that "Wholeness is shattered by the chaos of adolescence. Girls become fragmented, their selves split into mysterious contradictions."[00,260] Development shifts from the first and third functions to the natural second and fourth. In these functions, contests arise between authentic versus counterfeit influences from higher and lower CMs; contests of essential morality against self serving, of judgement and choice against possession states, of responsive feeling against unchecked emotional ties, of equanimity against casualness, of openness to higher purpose against pleasure and achievement for self. It is in this destabilising time of adolescent breakdown that the risk of catastrophe can lead to a court appearance, and subjection to uncomprehending court proceedings.

Analysing sexuality more closely, Whiteman believed "it is no exaggeration to say that it is impossible to understand much that is essential concerning sexuality and some other aspects of psychology without some reliable knowledge (or better, experience) regarding what I am here calling the 'control centres' in any human body, whether physical or non-physical."[00,204] He saw this as a "complex and confusing subject of 'control centres' (*chakras* in Indian tradition, also referred to by St Paul in Eph.4:16)."[06,120] The centres are not located in the physical body although they influence it; they "could not possibly be recorded by physical apparatus—they are mental impressions, or rather, archetypal ones somehow individualised in each person's physical body and also each individual's spiritual body."[00,205]

On the basis of his experience, and largely in agreement with Indian tradition, Whiteman recognised seven such centres, figured below, with functions (Chapter Ten) attributed to every centre.[06,121] The upper pair could represent brain lateralisation, the left half generally male (analytic, W) and the right generally female (synthetic, N). A single throat centre E initiates "breathing (spirit), speech (ideas), and hunger and thirst (in the spirit, for the Good and Right)." The upper and lower cycles overlap at the heart centre, S. Below the heart centre, a navel centre E can be seen as "controlling the growth of the (spiritual or physical) body, especially in birth and death."[06,121] Below are the male and female sexual functions, W and N.

W N
E
------S ------
E
W N

A physical body that is either male or female will be able to be expressive of both male and female functions on account of bisexual equipment of the lower non-physical control centres. One is related to male sex-instinct, the other he called the "birth centre" (*mūlādhāra)*. The two centres "obviously relate to the third and fourth creative functions, respectively, in their bodily working."[00,206]

The three "foundational skills" (Chapter Sixteen) are represented in the upper section as Active Recollection (male, W), Continuous Recollection (female, N) and Obedience (S).[86a,174;06,120] Whiteman emphasised that the control centres are not locatable in the physical body beyond impressions of locality, and "the wholeness of N is expansively *above,* and not actually in the brain; and an open responsive judgement (S) presupposes it."[06,121] W powers may give an impression "as if located in the forehead".[06,186]

The upper and lower cycles with the seven chief control centres were tabulated in a form allocating the upper cycle to the "first principle" of essence, and the lower cycle to the principle of existence (Chapters Eight and Ten), overlapping at the heart centre[93,246].

Noetic/essence/mind	**Substantial/existence/body**
supernal N	heart S
head W	navel (solar) E
throat E	maleness W
heart S	femaleness N

The different centres, and the division into separate but overlapping noetic and substantial sections, was seen in the context of shared functions, not fragmentation. "Our patterning on the 'Body of Archetypal Man-Woman' then results in all functions and all control centres contributing to our spiritual life."[93,249] Yet in his view, a special guiding relationship existed between the feminine supernal and birth centres: "no centre can function in a spiritually released way unless the supernal centre (Continuous Recollection, removing all conditioning) and the foundational centre (security in abandonment of self) have been stabilised, either on occasion and in an auxiliary way, or habitually."[00,208] Therefore the power of stabilising Continuous Recollection is principal in female core identity, but auxiliary in the male.[00,296] As noted earlier, in the male "the disposition to Obedience" was also considered to be "auxiliary and occasional."[00,207]

Consequently, "the Divine Source, when outwardly seen as if male, must nevertheless have female functioning as if behind or logically prior to it (the Great Mother, or 'virgin mother')."[00,296] He suggested, "This may account for the fact that so often in the ancient world God was said to be born from a virgin mother (meaning that the *knowledge* of God was born that way)."[00,208]

He saw "grave dangers, however, for anyone with male core identity setting out to gain supposedly mystical experience by self-directed efforts to force attention onto the birth centre, as in what is known as Kundalini Yoga." This was considered to be dangerously pathological and possibly antisocial "without practised Recollection, loving orientation to the Divine Source, male-female communion, or ability to sustain a spiritual life and meetings with non-physical beings in the higher worlds."[00,208]

Yet in ordinary physical life, "There is a continually shifting use of the functions, with hardly any recognition of their male or female character, or consequently of their principal or auxiliary character. Nevertheless, a sorting out of the motivations, with rejection of what is false, occurs in contests, especially during puberty, when the sexual functioning of the two lowest control centres forces attention and attempts at control."[00,207] This brings a focus of attention back to the two murder cases. Puberty is, as noted above, a destabilising time with the potential of adolescent breakdown, leading to the risk of catastrophe, a court appearance and subjection to uncomprehending court proceedings.

CHAPTER EIGHTEEN

CRIME AND PUNISHMENT

The inner personal complexity outlined in the last five chapters received virtually no attention in the two murder trials. Marlene, accused in the "scissors murder" of 1974 (Chapter Twelve), was found guilty with no extenuating circumstances by a trial judge. An appeal was granted, and in the hearing for appeal three possible explanations for the crime were debated in court: that she had an antisocial, psychopathic personality; that her personality had schizoid characteristics; that she had acted with deliberate evil intent.

Witness statements certainly did not support the idea of an antisocial psychopathic personality up to age 18, as noted in Chapter Twelve. In the court proceedings Whiteman recorded that "I have been unable to find any reference, by psychiatrists or legal men on either side, to dissociation, depersonalisation, derealisation, possession states, adolescent breakdown or insistent directive."[00,78] Happening in the early 1970s, the malign side-effects of Valium were not understood. Psychiatric opinion focused on the conventional good-bad-mad options of diagnosis, while the actual state of mind of the individual was sidelined. In this chapter Whiteman's analysis and its implications will be examined using the concepts of the previous chapters.

Whiteman compared the inner revolutionary events in girls of that age with Simone de Beauvoir's idea of "erotic transcendence" and the seeking for "at least the suggestion of a God-image in some male of this world."[00,351] When that happens, "she may pass into some stabilised degree of dissociation, with derealisation, and thus a clouding over of all social values incompatible with what the 'God' requires."[00,353] He cited the case of Myra Hindley of the Moors Murders, who despite having been of good character had also fallen hopelessly under the spell of a man.

The man in Marlene's case took advantage of sexual opportunity yet treated her callously. Whiteman considered three possible avenues for healing or escape for girls in a similar situation: getting away physically, suicide, or taking on "a stabilised state of depersonalisation, with loss of moral values at the root of the conflict."[00,356] The IC, he adds, is then either a dissociated spectator of events, or else falls dreamlike under influence foreign to the normal IC, influence which in Marlene's case was the disposing of the man's wife. Whiteman interpreted the foreign source of this influence as a CM "set on solving the problem in its own way" by imposing an insistent directive to a course of action which was not clearly understood. The course was nevertheless "felt to be morally or providentially necessary."[00,362]

Massive doses of Valium supplied by the man merely worsened the loss of core identity control. Marlene's state, Whiteman thought, "could be understood as one of possession, this remaining unrecognised because the lower CM has identified with the social situation and skills of the IC while retaining its own screening of moral values."[00,365] At times she returned to relative normality, yet "Evidently there was the kind of 'split mind' that occurs with insistent directives," he believed. "On the one hand, the individual normally in charge (IC) reacts and behaves with normal reason and morality; on the other hand, and fluctuatingly (presumably as the possessing co-mind comes and goes), what is an insistent urge to carry through a dreamlike plan takes hold in such a way that the IC, through the dreamlike character of the state then, cannot assess it rationally or morally."[00,371]

This interpretation raises special questions about responsibility and guilt that never entered the court proceedings. Yet they are core questions in the study and the exercise of spirituality or mysticism. Before pursuing this further, the case of Angela, Barry and satanic possession needs to be reviewed.

In 1992, on the morning before the murder of Angela's mother, Angela was reported to have told Barry to keep away from her because "something evil and bad's going to happen."[00,448] As noted in Chapter Twelve, both denied carrying out the murder voluntarily, and the task of the prosecution was to prove guilt by dismissing the idea of a possession state. The Prosecutor took the plea of possession as a "clever hoax", leading the Judge to declare Angela an "out and out liar", and Barry's evidence was declared "incompehensible".[00,495]

The prosecution called only one expert witness, a forensic psychiatrist who argued in favour of emotional arousal rather than dissociation. In a long discussion Whiteman showed the expert to have had an inadequate concept of dissociation. He ignored the possibility that the accused "were in a possession state and subject to an insistent directive on the day of the tragedy."[00,509] The Judge was not intellectually equipped to consider this possibility, and was misled by blatant lying on the part of the satanist woman when she was called as a witness.

Apart from showing the need for insightful assessment in such cases, Whiteman was concerned with the question of how such a case should be dealt with, given the gravity of what had happened. Central are the questions of free will, responsibility and punishment: problems of justice in the world as seen from a mystical view.

He saw the "problem of free will" to vary whether it is seen from a scientific, juridical, theological or ethical viewpoint. In scientific work he did not doubt that free will can be seen to operate in the investigating scientist, provided subjective as well as objective factors are acknowledged in his observations. Juridical assessment of free will came under his constant criticism of being unable or unwilling to take account of subtle psychological and psychic factors affecting an accused's free will; the attempted "solution" of the juridical problem by ignoring these factors "stands out as scientifically and ethically inadmissible."[00,530] The theological problem of free will was seen to centre round an intractable difficulty: "If an individual is able to make an effective choice which could be contrary to the determination of the Divine Providence, it would appear that God's power is limited by man's."[00,531]

As to the ethical problem of free will, "all that need be accepted is that, while our response to any situation in the physical life depends on both the providential script and our present disposition and skills, our disposition and skills change in the process of making a choice from the *higher level* of our core identity... This is how one develops spiritually."[00.531] He saw this as a "two-level process"; for example, in performing a memorised piece of music; the memory "script" is largely set at the lower level, but the choice of how the piece should be performed is made according to one's disposition at a higher controlling level. At the higher level (and our inmost core identity) "we are watching Wisdom; and such watching of Wisdom is precisely what Freedom means."[00,531] In unwisdom we have no "core-choice" and so are not free. The question of a "script" is discussed further in the next chapter.

In determining guilt, he listed four criteria to take into account during or before a crime: deliberate intent or purpose; adequate comprehension and moral evaluation; possible recklessness at the time; contribution to the happening due to negligence. He argued that if comprehension and evaluation are inadequate due to a dreamlike depersonalised state or to an insistent directive, then "one should not be regarded as responsible and therefore 'guilty' in law."[00,536] But even then, recklessness and negligence could justify some degree of a finding of guilt. These characteristics and the problems of free will have a bearing on what could be regarded as a mystical evaluation of guilt regarding the two murders.

Marlene, in Whiteman's view, presented an extreme depersonalised state and insistent directive. Despite the stress of her fraught relationship, deliberate suicidal or murderous action was countered by her habitual gentleness and moral control. However, "lower CMs fixated in the drive [to resolve her problem through murder] were thus impelled to devise an escape (as in a dream) which Marlene could not evaluate as real, but nevertheless would dutifully accede to as providing a God-given relief. Suicide being ruled out, relieving action was worked out (by the lower CMs) in a fantasy of murder... But the CMs fixated in their subconscious drive still demanded abatement in physical action by a 'dark agent'."[00,78]

He argued that she did not show negligence, in that she had communicated with her idol's wife, and in her state—worsened by innocent overdosing of Valium—it was in fantasy that she had the idea of "getting rid" of the wife. It might seem reckless to have taken her accomplice with her on the final and fateful visit, but Whiteman believed she did not take active part in the killing. He would therefore have argued: not legally guilty of murder with intent. But he was prevented from making his case in court as he was not counted as an expert witness from one of the relevant professions.

Whiteman believed her accomplice was "unquestionably subject to psychic infection along with verbal persuasion by Marlene in her drug-induced depersonalisation, aggravated by being sprayed in the face by the victim. This put him into "the category of recklessness in imagined self-defence."[00,538] Presumably Whiteman would have argued guilty of manslaughter but with extenuating circumstances.

Angela's talk of killing her mother was for Whiteman "the familiar catharsis of an adolescent whose burgeoning new identity seems agonisingly threatened by deeds of the loved and normally loving one." Hurting her mother was declared to be hurtful to herself, "Thus she could not have had the comprehension and evaluation of the fact that murder was taking place."[00,538] Therefore the first two grounds for guilt, deliberate intent and adequate comprehension, were not seen by him to apply. Recklessness and negligence were also not found to be factors in the possessed and derealised state she was in. Evidently, therefore, he would have argued not guilty of murder with intent in Angela's case.

With Barry, Whiteman believed "it was quite evidently an egodystonic possession state that dominated Barry at the height of the killing, for he struggled hard against the possessing force, only for it to become stronger."[00,537] The contest between harmful impulse and good character led to derealisation. Barry stated: "It was like a dream ... things would just happen; Time and space were in a mess; I myself had not killed anyone."[00,486] Whiteman found it difficult to see how Barry could be charged with recklessness or negligence. There was only an appearance of guilt, no true *mens rea,* a guilty or legally answerable mind.

He made the general point that in murder cases there seems to be a fixed idea in the legal and public mind that the more bizarre and horrifying the killing, the more certain it is the deliberate work of a permanently evil character. Yet he believed that this type of killing is an indication of precisely the opposite, namely a possession-state episode brought on by stress and/or drugs. "A sane murderer does not behave in a bizarrely irrational way."[00.539] There may be disagreement on this issue, but it does emphasise the point made constantly by him that greatest understanding and skill is required in making fair assessment of intrusions into the house of personality, often leading to conflicts and contests of worldly and otherworldly factions.

He did not feel called upon to deal in detail with the question of how to "make the punishment fit the crime," if punishment or correction were needed. He maintained that "True justice lies in an essential morality above the 'rough and ready' formalisms which attempt to deal effectively with the social troubles in this world. And the inflicting of pain on another person is never justified, otherwise than with a view to correction, guided by a genuine sense of wise but regretful necessity, and with as much loving support as we are capable of giving."[00,540] He conceded that this fails to work in practice because "expedience carries weight in this world over and above that of essential morality."[00,542]

The words "guilt" and "punishment", along with other related words such as "sin" and "repentance", were not favoured by him because of their ambiguity and loose usage. He argued that the Christian interpretation of "sin" has an emotive flavour that was absent in the writings of St Paul, where the Greek origin of "sin" is *hamartía,* meaning to "miss the mark, or fail of one's purpose."[00,548] Therefore it should not incur "guilt" in the normal emotion-laden sense of the term. He regarded "sin" and "guilt" as "smear-words" because they are used "with the implication that an emotive rejection is to be categorically attached to a certain person whose body performed a certain deed, regardless of possible extenuating or absolving factors, such as social background, drug influence, and state of mind of the person at the time."[00,550] Being smear-words, they are liable not to be carefully weighed and chosen.

The same is true of the words "repentance" and "punishment". In the New Testament, the word "repentance" is a rendering of the Greek *metanoia,* which means a change in heart or mind regarding purpose. "The change," Whiteman wrote, "is naturally from some less satisfying state of life to a better one, but there need be no suggestion of guilt." Rather it is "marked by a settled change from rule by the self's pleasure and pain to acknowledgement of the prior validity of orientation to the right and good."[00,522] This was John the Baptist's "baptism of repentance", for overcoming *hamartía*. Punishment, then, needs to take note of this smear-free and unemotional, objective understanding, since "the administration of stress or handicap for another person can be helpful or damaging, and perhaps both at once in different ways."[00,553] The aim should be *metanoia,* ultimately the release of the individual from his samsaric house of personality. This will be discussed in the next chapter.

CHAPTER NINETEEN

DESTINY AND RELEASE

At the beginning of this Part Two on the house of personality, it was asked, is the individual confined, bound to his/her samsaric house, or is there a way of release? The connection between release, space and time was discussed particularly in Chapter Nine; in this chapter the focus is more on the individual, on what, in his *Four Quartets,* T. S. Eliot (1969) called liberation from

> Attachment to self and to things and to persons,
> ..... not less of love but expanding
> Of love beyond desire, and so liberation
> From the future as well as the past.

For Eliot the desirable outcome of release is positive and creative, not mere escape. Attachment and fixation being overcome, "the mind," as Whiteman wrote, "opens out to universality."[67,395] "The final release," he recognised, "is a confirmed transcendence of the attachments of 'self' (*atta*), with all its rigidities of thinking and belief, its physicalistic presuppositions and 'cloak of ideas', in such a way that all four contemplations [*jhānas*] can be freely exercised from a higher wisdom in the light of what is absolutely and finally 'uncompounded'."[67,110]

Both Eliot and Whiteman picked out attachment to self as a prime hurdle in achieving release and development. Self is a dominant concern in the literature on mysticism, and for that matter also in science if one considers statements such as Heisenberg's remark quoted in Chapter Two that the sharp separation between the world and the I is impossible. This is recognition of the second power of *māyā,* enclosure. It revolves round the idea of self and its complex isolating and attaching powers. One of the leading principles in early Buddhism is that "all pure purposes or conceptualisings (*dhammas*) are selfless (*anatta*),"[93,45] which implies that it is an error to attach self-reference to such things, because then there is little chance of experience being objectively perceived and assessed.[93,188] As a result one is less able to learn from experience and become detached from it, as discussed in Chapter Twenty-one. Whiteman asserted that "*anatta* must come to rule our lives entirely."[06,88]

Yet an individual in charge (IC) is prone to the error of self-involvement, usually misled by self-seeking co-minds (CMs), so "the cardinal principle of release" is to objectify CMs or CM influence, and so deprive the unruly mob within the house of personality of their attaching effect.[93,274] In this endeavour the fourfold creative cycle (Chapter Ten) should come into play, requiring enthusiasm or energy, guided by wisdom and judgement, followed by creative action, then reflection and build-up for further enthusiasm towards "ultimate release (*mokṣa*) through conformation with a transcending and immanent Divine Humanity."[93,12] *Mokṣa* signifies release altogether from the house built on the samsaric wheel of life, beyond the domination of self.

This pilgrim's progress, with its cycles, conflicts and contests, is, according to mystical tradition, staged in a life-plan for every individual within an overall world-plan. The previous chapter noted the idea of a script or score from which we play out our life, and it leads to a question: is all the world really a set script and stage, and we are merely players who have little freedom beyond playing our parts well or badly? Whiteman gave prominence to a passage from *Hamlet:* "There's a divinity that shapes our ends/Rough-hew them how we will." He cited Nostradamus's prediction of the arrest of Louis XVI at Varennes and the Reign of Terror, 240 years after the prediction was made;[00,99] does it indicate a life-plan that the King could not escape? Are we the subjects of preordained destiny, with a purpose no better than a poor player soon heard no more, signifying nothing?

Bearing in mind the idea of Purpose, Means, End, the hypostases of essence (Chapter Ten), Whiteman saw a "universal Purpose" to be "a time-spread potentiality leading eventually, in a directly causal way, to the experience of living beings in a developing environment, such experience being finally actualised according to each individual's state of life, disposition and mental skills."[00,98] On this rather shifting stage the player's performance is scripted within universal Means, which involve "characteristics of the developing individual."[00,99] Yet Means can affect the development of potentiality; the stage to some extent responds to the performance of the players. A kind of feedback was envisaged, in that an individual's performance can start the beginning of another cycle of Purpose which, being potential, can affect one or any number of other people. So at the End "all individuals can be united, with varying degrees of empathy, in a larger Purpose for all Humanity."[00,99]

This can be seen as a "world-plan" that expands continually as stages in it are successively actualised; and in a limited way the individual could affect the world-plan and achieve, for example, intervention in a precognised happening as discussed in Chapter Two. This possibility rests on there being "less important and perhaps provisional stages in the World-Plan, as well as room for innumerable details that would not be determined till actualisation occurs." He added, "The individual's choice is thus not bound by a rigidly predetermined Plan, but may be very largely directed by his present disposition or specific needs."[00,100]

The Plan was seen to provide "the framework for individual choices and guidance"; otherwise, "individuals, when required to 'fill in the gaps' such as are always available to them in the Plan, could not form any sure judgement as to what is ultimately wise and what is ultimately unwise. Wisdom would be overwhelmed by chance."[93,116] He therefore called it a "Divine Plan for Humanity", or "Destiny": "To go against it, wholly unconditioned as it is, is to condition oneself to unwisdom; and to conform to the Plan, obediently and not self-centredly, when we see what is absolutely and universally beyond conditioning, is to behave wisely."[93,127] The conditioned (namely, "disposed to act in an unwise or unskilled way through lack of reflection") and the unconditioned is discussed in Chapter Twenty-one in connection with *karma* and disposition.

In this "world-plan" Whiteman thought that "the principle aim in such Purpose is development of the individual through contests by which spiritual skills are cultivated."[00,99] The releasing spiritual skills of recollection, faith and obedience, were discussed in Chapter Sixteen. The catastrophic events in the lives of the people involved in the two killings need to be seen in this light, so Whiteman maintained. Among it all were opportunities for spiritual growth, even through exposure to depersonalisation and derealisation, "thus being introduced to the foundational state of stabilised or habitual non-attachment which is necessary for an eventual mystical transformation."[00,426]

Such a transformation goes beyond the requirements of the definition of mysticism noted in Chapter One. Yet the definition could well be enlarged to include "a certain *transformation* of bodily and mental life, described as 'release' in Buddhism and 'non-attachment' in the Indian scriptures."[06,177]

Nevertheless, when considering why people are drawn into life-plans that have horrific endings, Whiteman conceded that in our present state of life we are not wholly fitted to find answers. "To believe the World Plan to be reasonable and just in such a case is necessarily a matter of *faith* and not of *information*; and we should not expect all tests of faith to be done away with."[00,36] Yet following this caution he added that we may find grounds for belief in a World Plan "as the picture of spiritual development for individuals becomes clearer."[00,36] One way this picture can become clearer is to survey its history within the mystical tradition. This is attempted in Part Three.

Part III

Spiritual Development

CHAPTER TWENTY

SPIRIT, LETTER, AND MYSTICAL EXPERIENCE

How far can spiritual method and development be taught, or even usefully discussed? It seems to be a feature of civilisation that searches are made for some kind of answer; this is the spirit of civilised thinking, even if the letter—the resulting theory and dogma—is inclined to smother it. "The letter killeth, but the spirit giveth life," as St Paul wrote to the Corinthians. This Part will join Whiteman's search for this spirit, often concealed by a "cloak of ideas", as Husserl termed it. Consistent with the phenomenological approach, Whiteman wrote that "nothing is presupposed" in his synthesis of science, mysticism and psychical research. "The mind is opened to a certain kind of experience, which we have been able to develop in our lives, while being also able to understand the nature and implications of such experience and to develop it."[00,vii]

Experience was seen even to enter the root of theology. For Whiteman the term simply meant "the application of the phenomenological method to our awareness of the 'Divine', in the wholeness and detail of its meaning, omnipresent in our lives if only we adopt the right *releasing* and *uplifting* attitude in life."[06,vii] So, "the word 'God' (*Theos*) must be taken to stand for the Archetypal Reason in all."[06,64]

With placing experience of spirit above the letter, his critical analysis of traditional texts tends to differ radically from conventional treatments. In the first place this is a result of his reading Vedic, Sanskrit, Pali, Hebrew and Greek scriptures in the original, evaluated against his own mystical experience. Thus he wrote, "In existing English translations and scholarly studies of the hymns of the Rig Veda I have been unable to detect any sign of mystical knowledge. The words seem to have been chosen so as to suggest nature worship, a fertility cult, or a mere instinct to worship imagined divine beings or objects and to carry out rituals concerning them."[00,293] His strongly held view was: "The two great errors made by traditional translations of the Rig Vedic hymns thus seem to be, firstly, in assuming that all references are to experience at the physical level, and secondly, avoiding any suggestion of poetry, with its subtle devices suggestive of time-like-space recollection."[00,294]

Translations of ancient scriptures have, he regretted, "almost invariably been made by scholars lacking mystical experience. The words chosen then, when a passage refers to experience at a mystical level or describes features of the 'mystical way', almost inevitably suggest a non-mystical meaning, quite false or irrelevant in actual experience."[06,viii] This is a reflection of a general lack—if not suppression—of non-physical understanding or experience in the "educated world", as Sidgwick found so deplorable (Chapter Eleven). Whiteman echoed Sidgwick in writing, "the facts concerning mystical states of knowledge and life seem to remain, even for the majority of educated people, a closed book."

Nearly all the quotations from ancient texts included in his work "have been newly translated from the original languages, hopefully so as to suggest something mystically correct."[06,64] A particular feature of his writing was an unusually deep knowledge of Pali, the language close to the dialect in which Gotama Buddha generally spoke, and in which the Buddhist canon was first collated.[83,8;06,160] After this time, Buddhist texts were transferred to Sanskrit, generally without appreciating crucial differences between seemingly equivalent words in Pali and Sanskrit.[83,8] As Pali is not widely studied, commentators on Buddhism generally base their opinions on Sanskrit versions, usually on the letter of later Mahāyāna divergences from the original Pali cannon. The early canon is then termed Hīnayāna, from *hīna,* meaning inferior, faulty, to be rejected (letter of 2/2/1999). One could be forgiven for thinking, "What a mess."

A second difference from conventional treatment is that Whiteman restricted his inquiry "to writings reasonably taken to refer to first-hand mystical experience in history up to about 120 CE. This is because there is good reason to believe that, after that date, the earlier and genuine accounts of mystical experience have been built on by theologians and others who had not been granted such experience."[06,xiii] This led to losing sight of "the mystical harmony of the scriptures in the very diverse languages of the past."[06,vii]

Whiteman's definition of mysticism was noted in Chapter One. This definition appeared at the beginning of his first book, *The Mystical Life* (1961), and naturally his presentation became elaborated. In *The Dynamics of Spiritual Development* (2000), when discussing the "archetypal structurings" of the good, the true, the beautiful, he wrote, "We might even define the 'mystical' as 'what is recognised as manifesting these three kinds of archetypal valuation, and not conformed to this world.'"[06,3] In his last book, *Universal Theology and Life in Other Worlds* (2006), a further elaboration of the definition appeared under a heading, What is "Mystical"? "The mystical way could be said to begin when our life is habitually ruled by a spontaneous and steady direction to the Right and Good (instead of self-advantage, desire, or mere custom). It may lead to becoming prepared to accept that spiritual worlds (and this world too) can exist with a kind of Reality (beauty, life, power) not limited to the seeming 'shell' of matter 'out there'."[06,177] Along with this, as already noted (Chapter Nineteen), goes "a certain *transformation* of bodily and mental life, described as 'release' in Buddhism and 'non-attachment' in the Indian scriptures. In Christianity, attainment of such release is prompted by the instruction 'let go and let God' (that is, watch for and faithfully obey the Right and Good)."[06,177]

When spirit gains over letter, there may come "a transcending awareness of the *Source* of the Right and Good, perhaps even *seen* in certain remarkable ways—referred to in all major religions."[06,177] Use of the phrase "the Source of the Right and Good" in this late work of 2006 is consistent with his early definition of mysticism in 1961; there was no mention of a particular god or deity. Later in the same book he expanded the meaning of the term mystical "to distinguish those higher experiences in which there is a clear and genuine recognition of the Source."[61,119] By 'source' was meant "the universal Source of Goodness and Wisdom."[61,17] The word "God" did appear in his 1961 *The Mystical Life,* but he was cautious about use of the term, as when "the name 'God' is associated in the mind of the subject with theological beliefs which are not warranted by anything in the experience itself, and these beliefs are then misleadingly embodied in the description of the experience."[61,119] He maintained that scientific mysticism needs to be critical of religious beliefs and terms not directly warranted by experience, also the "letter" where "theoretical interpretation has been arbitrarily imposed."[61,119]

In his last book, *Universal Theology* (2006), instead of "God" there appear phrases such as "Archetypal Reason in all"[06,64], "source of Right and Good"[06, 177], "supreme Source and Universality of higher values"[06,54].

Awareness of the Source of Right and Good will, Whiteman wrote, "go along with a re-making of one's bodily and mental being, and a revelation of one's secret *core identity,* as an ideal 'Son' or 'Daughter', hidden within all that is physical."[00,177] With this development comes the recognition of a corporate structure of personality (Chapter Thirteen) in which the core identity is usually concealed. The following chapters will have much to do with the failure of translators and commentators to be aware of this corporate structure, and so offer fantasies based on physical interpretations of what is not physical. In the Old Testament, Whiteman saw "the people" to be a code for the constituents of personality, and "foreigners" represent influences not belonging to the settled corporate structure. In the New Testament, the "Son" was a code for the revealed core identity (Chapters Twenty-four and Twenty-five).

Commentators more readily and accurately recognise release and upliftment as "a very fundamental mystical teaching."[06,179] Along with this is seen the necessity for confronting self and its propensity for attachments. But the underlying processes of development tend to be coded by spiritual teachers, and so receive less attention by commentators. Notable are stress and contests experienced in spiritual growth, and the cycles involved in development. The fourfold creative cycle and the associated skills will appear frequently in the next few chapters, but rarely in the work of commentators; yet these are the mills of spirituality.

Islam was not covered by Whiteman. The "great obstacle", he saw, was that "I do not know Arabic, and have no time to become proficient in it."[06,xiv] (This was written shortly before his death.) Experience led him to write, "Knowing how other sacred texts have been distorted in translation, I cannot make an exception in this case."

Not a good commentary on the quality of the "letter".

CHAPTER TWENTY-ONE

THE YOGA SUTRAS

Whiteman had particular interest in the oldest religious tradition still widely in practice today, one that can be traced back more than 1000 years BCE to northern India. The hymns of the Rig Veda date from this time, followed by the Upanishads started around 800 BCE, and then the Buddhist reformation around 500 BCE. A unique crystallisation of Indian spiritual method in the form of aphorisms or *sutras* occurred from about 200 BCE to 50 CE when the aphorisms were assembled in four books, the Yoga Sutras. They are conventionally attributed to the grammarian Patanjali (*c.* 150 BCE), but they reach far beyond the time and work of one man.[93,4] Whiteman believed the *sutras* to "bear the hallmark of firsthand experience correctly understood; they have nothing of the character of philosophical conjectures."[93,3] He published an extended treatment of them in 1993 as *Aphorisms on Spiritual Method,* and this chapter explores his treatment as much as the *sutras* themselves.

The aphorisms, written in Sanskrit, may be open to very different interpretations on account of the epigrammatic *sutra* style, its lack of prepositions and pronouns, and the inclusion of many-word compounds. The thrust of Whiteman's interpretations is indicated by the subtitle of his book: *in the light of mystical experience.* The non-mystical, physical interpretation of passages in the *sutras*, often encountered in English translations, were rigorously criticised in this radically different treatment. As he pointed out, the conventional Indian interpretation of the *sutras* from about 250 CE onwards was heavily influenced by the development of a non-mystical Sāṅkhya philosophy, to which English writers without much mystical (or even psychic) experience largely referred. Whiteman's interpretation returns to the Vedas and Upanishads, and to early Buddhism which, he showed, was incorporated in the Sutras. His expert knowledge of the Pali Buddhist discourses as well as the Vedas and Upanishads makes his translation exceptional; English translations of the Sutras have conventionally been made within the later Hindu tradition, which regards Buddhist teaching as heretical, but Whiteman emphasised that "From the dialogues of the Buddha, technical terms in abundance have been taken over in the Yoga Sutras, and these sometimes appear (Sanskritised) in what are virtually quotations of Buddhist teachings."[93,xiii]

His translations are also exceptional in being set in rigorous scientific thinking, to which he was a contributor. The book therefore becomes a synthesis of science, mysticism and psychical research in a way not found in other treatments of *"The Yoga Sutras of Patanjali"*. Twelve introductory chapters focus on the *sutras'* historical context, including a study of the Sanskrit language and the Pali influences. Each *sutra* was presented together with an interlinear word-by-word translation, accompanied by a carefully considered idiomatic translation with textual and interpretative commentaries. The English renderings tend to differ markedly from conventional editions on account of being viewed through Buddhist and his own mysticism.

Most of what has been covered in previous chapters of this book appears in the *sutras* in condensed and explicit form. The 51 *sutras* of Book 1 cover spiritual development from the practice of non-attachment through to the achievement of what Whiteman termed Continuous Recollection (Chapter Nine), and continue in Book 3 to "the ultimate goal in the Yoga Sutras", the "all-one-ness" of *kaivalya*.[93,29] In his translation, *yoga* is the checking of mental involvement (*sutra* 2) through stoppage and non-attachment. Stoppage is effectively the same as Husserl's finding that "by a suspension of attachments, and a 'stoppage' (*epoché*), we come to a 'unique form of consciousness', characterised by 'a certain refraining from judgement which is compatible with the unshakable because self-evidencing conviction of truth'."[67,108] (Chapter Nine.) Stoppage or checking "is by application and non-attachment" (*sutra* 12), yet *sutra* 22 asserts that beyond the characterisation of non-attachment is "something primordial", an "acknowledgement of the Divine Source (Puruṣa)" (*sutra* 16). It calls for Obedience (*sutra* 23, see Chapter Sixteen).

The word *puruṣa* has the normal meaning of "man", but from the Rig Veda through the Upanishads it developed successive connotations: divine personage, Divine Man, Divine Humanity, Divine Source, Īśvara or Lord, Supreme Goal,[93,28] so that it became to stand "for both the Archetypal Humanity and the creative Spirit which flows from it from a Divine Source, into everything that exists in created life."[93,168] This mystical meaning of *puruṣa* was taken up in the Yoga Sutras, although discarded in later Sāṅkhya philosophy and the interpretations of the Sutras based on it.[93,xvi] In authentic Eastern and Western mysticism, Whiteman wrote that the "Divine Source" is uniformly taught to be "perceived as the kernel of 'real life'"[86a,77] or "the great Source of Wisdom, which we see as a reality beyond the physical but freely available to us for the comfort of faith and the teaching of what is right and good."[86a,152]

It might seem to be fiddling with the "letter" to try to make distinctions between the connotations of *puruṣa*; nevertheless the successive meanings from "man" to "Divine Source" indicate a process of spiritualisation, rather like the progressive spiritualisation of "God" in the Old Testament. Essential to a "strictly mystical" meaning of *puruṣa* was the "recognition of a Divine Source of Wisdom, which, through our orientation of Faith and Obedience, removes all attachment and transforms us more and more perfectly into a 'harmony and likeness' (Book 3, *sutra* 56) patterned on a Divine Humanity."[93,62] This is consistent with Whiteman's definition of mysticism recorded in Chapter One, where no mention is made of a particular deity or god.

All this testifies to a reach of experience that scientific analysis has virtually no practice in handling, yet Whiteman's translations of the *sutras* show there was a rigorous treatment of the practicalities of striving for non-attachment. Non-attachment and stoppage must be practiced on a whole range of mental functions—warranted knowledge, mistaken knowledge, opinion, inattention, memory—until it becomes habitual (*sutra* 6). Non-attachment is dissipated by illness, stolidity, doubt, carelessness, laziness, intemperance, confusion, slow-wittedness and lack of control (*sutra* 30). To ward these off, a number of measures are listed: compassionate behaviour, regular breathing, engrossing activity, contemplation of the beautiful, reflection on dream; "For when anyone has entered the path of self-discipline, there is need for a power of control extending to all kinds of experience from the infinitesimally small to the supremely great." (S*utra* 41).

This is followed by *sutra* 43 which in Whiteman's translation corresponds with Husserl's standpoint: "Upon exercising complete purity of recollection achieved through stoppage, there arises a non-speculative state, which is as if void of subjectivity and illuminated by the meaning only." Whiteman pointed out that the *jhānas* (Chapter Nine) are also represented in this key *sutra* leading towards Active Recollection (*sutra* 46) and "serenity and grace in personal life" (*sutra* 47). *Sutra* 48 sides with Husserl again: "In that state, discernment carries with it assurance of self-evident truth." (Chapter Eight.)

Sutra 49 states that this kind of discernment "has the character of covering a domain of experience and knowledge different from that of discernments depending on testimony or inference." This again is aligned with Husserl's description of a special kind of detachment and non-speculative state that allows intelligible analysis and immediate seeing in a state of "alienation from regular life" (Chapter Eight). This "higher kind of discernment", which Whiteman regarded as Active Recollection, "holds in check all other habits" (*sutra* 50). By a checking even of that arises a state of Continuous Recollection (*sutra* 51).

The term Recollection, discussed in Chapter Nine, corresponds to the word *samādhi* (collection[00,172]) in the Sutras. Whiteman considered English renderings of *samādhi* as mindfulness, concentration or absorption to be misconceived; he followed the early Buddhist treatment, where *samādhi* (*sati* in Pali) includes the four *jhānas* (Chapter Nine) and the *abhiññā* over-knowledges (Chapters Twelve and Twenty-one). Active Recollection (the second *jhāna*) and Continuous Recollection (the third *jhāna*) were translated from the *sutras* as "collectedness with seed" and "collectedness without seed" respectively; "without seed" signifies that even the checking (*nirodha*) of habitual reactions in Active Recollection is finally checked in Continuous Recollection to achieve "a unique form of consciousness", as Husserl termed it.

Whiteman followed Book 1 immediately with Book 3 in his section on the text in idiomatic English; it seems to follow Book 1 more directly than does Book 2 and "may have come from the same hand."[93,4] These *sutras* focus on *saṃyama,* which Whiteman translated as Recollective Control, namely attentiveness, leading to narrowing the field of attention to some detail, and to "the crucial movement of release through 'stoppage', or holding back of the current of time on the conceptuality of what is perceived."[93,223] This is clearly one with Husserl's phenomenology (Chapter Eight); it leads to "objectification of habits or tendencies" and the releasing of attachment to self through knowledge of previous "personality phases", as stated in *sutra* 18. The word *jāti* appearing here is commonly given a physical meaning of "birth" in a physical sense, which obscures its spiritual meaning as the occurrence of personality-phases that are impediments to spiritual development; personality-phases governed by attachment have constantly to be checked (Chapter Thirteen). Among such "conditioned habits or tendencies" Whiteman included "automatic instinctual tendencies, wishful thinking, desires, formulas of thought, catchwords, automatic personal identifications by thoughts such as 'I am like that; that is mine, I want that', and so on."[93,236]

The resulting state of release and dissociation, treated in *sutra* 21, also tends to be given a physical interpretation in the general literature: the yogi's body is supposed to disappear, rather than there being an "inner placing" or separation (Chapter Four), disjoining oneself from the physical body through Recollective Control. Similarly, in *sutra* 39, reference is to entering "another body made of mind", an obvious reference to separation;

yet commentators unfamiliar with separation suppose reference is being made to the magical power of a yogi to enter a dead body or another living physical body. The phenomenon of mediate identification, discussed in Chapter Fourteen, could be relevant to such cases. Separation is said in Whiteman's rendering of *sutra* 39 to be achieved "By relaxation of the causes of binding attachment to the physical body, and with full knowledge of the manner of going forth."

Whiteman's rendering of *sutra* 51 reads that, with the habitual achievement of tertiary separation (Chapter Four), "all seeds of transgression being in the process destroyed, All-oneness [*kaivalya*] described as the 'unitive life' will follow in due course." This leads to the last *sutra* of Book 3, number 56, "All-oneness consists of purity, equipoise between the individual and the Divine Man (*Puruṣa*)." Whiteman stated, "orientation to the supreme Puruṣa is held to be the prime cause of release and the 'all-in-oneness' (*kaivalya*) which is the ultimate goal of the Yoga Sutras."[93,29] The state of "all-in-one" was portrayed in a letter to Dr A. Lichtigfeld (10/9/91) as being "like melodies in music, where each note has a character and effect only through its relation to the whole." He also described it as one of complete release in which "all phenomena and existences are known as if indwelt by the One Divine Source of the Lord (Īsa Upanishad, verse 1): and the one individual is potentially present (and knowable in that way) in all that exists, great or small."[93,26]

The term *kaivalya* is derived from the word *kevala,* alone; "the individual is 'all-one' or 'alone' (*kevala*) in the sense that the unitary perfection of that life has raised it above all disruptive attachments and made it fit for participation in the primordial divine harmony."[93,21]

Sutras 30 to 33 of Book 3 comment on the *cakras,* often spelt *chakra*, which Whiteman rendered as control-centres. They were discussed in Chapter Seventeen on sexuality, as forming two interlocking creative cycles operating in the upper and lower halves of the non-physical body, a male-female pair in each. Only four of the Yoga Sutras touch on the *chakras. Sutra* 30 was translated as, "Knowledge of the organisation of the spiritual body comes by Recollective Control in respect of the control-centre in the navel region."[93,245] Whiteman noted that knowledge of the navel or solar centre, "flooding out a transcending warmth and energy, is sometimes (perhaps always) the first to become vividly objective." He suggested that Pascal's "Fire" seemed to have come first in this centre, and "energised a mystical disclosure proper to the heart or supernal centre."[93,246]

Book 4 of the Yoga Sutras also considers spiritual development leading to *kaivalya.* Whiteman's interpretation of these *sutras* rests on his recognition of the corporate structure of personality (Chapter Thirteen); this makes his interpretation differ radically from conventional commentaries, which fail to perceive this structure. The main thrust of Book 4 is seen to be the discerning of contributory minds or co-minds, *nirmāṇa-cittāni,* and their clearance towards achieving *kaivalya. Sutras* 4 to 6 were noted in Chapter Thirteen to be "the cardinal principle of release".[93,274] *Sutras* 25 and 26 are then rendered as, "In the case of one who perceives the distinctness of co-mind influences in the personality, there is cessation of the dwelling on self. Then, indeed, having the full depth of detachment, the mind is borne forward to All-oneness [*kaivalya*]."[93,291]

Book 2 focuses on "taints" and attachments that hinder spiritual development, but seldom reaches great depth. *Sutra* 3 is rendered as, "Taints are classifiable as kinds of unwisdom, specifically self-centredness, drive to attachment, antipathy and malice, and clinging to ideas and incentives." Whiteman saw *sutra* 6 as developing the basic Buddhist teaching that "all *dhamma* (pure formative principles or purposes in life) are without self (*anatta*)," discussed in Chapters Eighteen and Twenty-one: *atta* is self with the tendency to attachment, regarded as a taint. He commented, "Widespread confusion concerning this *anatta* teaching has resulted from the English rendering of *atta* as 'soul'." Then *anatta* means "no-soul", which lends itself to a materialistic interpretation. Yet, "If the 'soul' is our enduring individuality, growing all the time in secret, then far from denying 'soul', it points to its existence, as beyond 'self'."[93,188]

Sutra 12, which contains the word *karma,* is rendered as "Possession of motives rooted in taints is to be comprehended in terms of personality-generation, either seen as due to co-mind influence, or unseen in that way, but knowable in its effects." As with the misunderstanding of *atta,* use in English of the word *karma* was criticised by him. Conventionally taken to mean "deeds" or "action", Whiteman pointed out that *karma* at the time the Yoga Sutras were being assembled meant "primarily a motivational tendency, and only rarely some specific motive which is approved and allowed to express itself ultimately in some bodily act."[93,173] Hence his translation of *karma* as "deed-attachment" as in Chapter Thirteen, namely a motivation, not a deed in itself. So "bad *karma*" would have meant "bad disposition or motivation", not "(the result of) bad action", as presently understood.

Whiteman devoted a whole chapter of *Aphorisms on Spiritual Method* to *karma* and to *akarma,* the absence or shaking off of *karma.* A difficult passage from the contemporaneous Bhagavad Gita was cited: "He who in *karma* can see *akarma,* and also in *akarma* can see *karma,* he is wise in all human affairs."[93,117] He concluded: "Motives in this world are almost always conditioned in some way. If then we are to have 'wisdom in human affairs', the conditioned motives in us and in other people must be recognised and held off—not repressed or escaped through fear, but intelligently allowed for and gently removed, so far as possible, by placing the Unconditioned and its equanimity against them. In other words, *karma* (conditioned motivation) and *akarma* (the Unconditioned) must be seen as a whole, neither being pushed quite out of sight. Each, when we focus on it, is seen to emerge from the other. The teaching of the Bhagavad Gita, in IV:18, has been found true and helpful."[93,127]

Book 2 continues with an analysis of attachment, in the sense of fusion of self with an object (*saṃyoga*). *Sutra* 17 is interpreted as: "The cause of the ill which is to be prevented is fusion of the perceiver with the object perceived." This leads into a discussion of objects and perception, with an overall aim of liberating the perceiver from attachment to experience of objects. Conventional treatments tend to get lost in a supposed philosophical problem of *sutra* 22: What happens to a physical object when we are no longer looking at it? (Recall the discussion in Chapter Two.) The answer conventionally ascribed to this *sutra* is that the object is "in common" with other people, but, as Whiteman points out, no-one might be looking at it, and so the problem remains. The *sutra* may be seen as itself coming close to the taint of clinging to ideas (*sutra* 3), and Whiteman comments, 'such a philosophical problem, suddenly introduced here, is irrelevant, and interrupts the logical flow of ideas."[93,201]

The analysis ends with *sutras* 24 and 25 which are fairly similar to *sutras* 25 and 26 of Book 4; Whiteman interprets them as: "The cause of fusion is the unwisdom of the perceiver, in being unable to distinguish 'self' [personality] from what is proper to the individual in charge. Habitual absence of fusion, through absence of that unwisdom, is therefore the required preventative of obsessive ill. That is the All-oneness of the perceiver."[93,203]

This leads to a discussion of the "limbs" of yoga through *sutras* 26 and 27, interpreted as: "Persevering dedication to detachment is the means for preventing future ill. In the case of one who attains to *kaivalya,* there is an ultimate ground of discernment and wisdom, developed in seven stages." Whiteman considered this to refer to the "seven limbs of enlightenment" in Buddhist teaching; eight "limbs" are in fact listed in *sutra* 29, including moral restraint, spiritual dedication, breath control (*prāṇāyāma*), contemplation (*dhyāna*) and collectedness (*samādhi*). *Sutra* 41 is interpreted as "Purity in one's inmost being consists in benevolence to all, the power of unitive uplift, mastery of the senses, and fitness for contemplation of spiritual things." Whiteman detected the four "creative functions" here (Chapter Ten), starting with East, benevolence.

Yoga is conventionally associated with bodily posture and breathing, a topic covered in *sutras* 46 to the end of Book 2. *Sutra* 50 is translated as "Breath control, as concerned with what is external, internal or transitorily arrested, is to be studied by attention to place, time or counting, in a prolonged way and with deep discernment." The usual rendering of arresting breath is "cessation", which, as Whiteman points out, is a sure way to create unease. "If one stops the flow of breath, one has ceased to be able to control it."[93,218] He commented that breath-control has three aspects: physiological, aimed at ease and correcting discomfort; psychological, aimed at tranquillity of mind; and "control by will, in wisdom and judgement." The control acts "by 'breaking into' [not stopping] or transcending the normal course of breathing."[93,218]

The Yoga Sutras can be regarded as a compendium of Indian thinking up to the start of the Christian Era, after which theorising and loss of mystical insight set in, as it did in the Christian tradition. The Sāṅkhya philosophy that emerged in India was characteristic of this trend. One of the important features of Whiteman's treatment was the study of the *sutras* taken at the time of their writing, largely the period of the great Maurya Empire that spanned most of the Indian subcontinent. During this period there was "a *rapprochement* between Buddhism (liberally supported by the emperor Aśoka, 274-232 BC) and a brahmanism in process of reform, also supported by him."[93,65] This evidently accounts for the easy admixture of Vedic, Upanishadic and Buddhist insights in the Yoga Sutras.

The only major feature of early Indian thinking missing from the Sutras is discussion of the *kalās* of the Upanishads, outlined in Chapter Ten. This might have been regarded as too theoretical. Perhaps for the same reason, less attention was given to the intricacies of potentiality, discussed in the next chapter. Yet when it comes to a treatise on spiritual development, the Yoga Sutras are without parallel, as is Whiteman's exploration of them.

CHAPTER TWENTY-TWO

BEFORE THE YOGA SUTRAS

In his *Universal Theology* (2006), Whiteman translated pieces from the Rig Veda and Upanishads that present Indian thinking before the time of the Yoga Sutras. Their subtle, poetic handling is largely lacking in the Sutras, which Whiteman saw more as a collection of "leading ideas on an important theme, in the manner of brief lecture-notes prepared by a teacher, and no doubt enlarged upon in later discussion."[93,3] Even though framed in a poetic and pictorial way, Whiteman showed Vedic and Upanishadic thinking led more directly than the Sutras to underlying issues such as potentiality and actualisation, the creative functions, and the hypostases of existence, discussed in Chapters Two and Ten. The thinking could be called "philosophical", but more universal than in professional philosophy today, and more poetically expressed.

Potentiality and actualisation were seen to present an indivisible whole. So when the author of the great Muṇḍaka Upanishad surveys a variety of natural things, it is not to provide a nature study but to remind us that "each has its higher meaning and life, integral with the one universal Humanity."[06,152] In these terms, to "see God in everything" seems less strange to contemporary thought. Whiteman wrote, "It is what presents itself, when one is released in the highest state, as the God-Source (*Brahma-yoni*),"or "Light of Lights" or "Divine Sun of true Light of Life" as declared in the Muṇḍaka Upanishad.[06,152] Section Four of this Upanishad, on "The Divine Source", starts with the Great Stride of Vishnu (Chapter Ten), "the three levels of causation, as they appear, for example, in quantum physics—purpose or boundary conditions, logical structuring, and manifestation upon occasion."[06,155]

> Manifested, constituted, secretly proceeding,
> this is the "Great Stride".
> So, this universe is caused to come forth.[06,149]

The Great Stride is also evident in translations from the Rig Veda in the *Hymn to Agni:*

I call on Agni, primordial domestic fire,
Divine and timely ministrant of human prayer,
Well versed in acts of sacrifice,
Thus bringing gifts beyond all price.[06,136]

Agni means fire; in this hymn it is protective and home-warming, and is likened to three kinds of priest, which correspond to the Great Stride: "creative purpose and motive, intelligible means, and fulfilling actuality."[06,138]

Rather similar to the quotation from the Bhagavad Gita concerning *karma* and *akarma,* noted in the previous chapter, is a verse from the Īsa Upanishad that Whiteman translated as:

He who knows knowing and not-knowing, both together,
having by not-knowing crossed through death,
he then, by knowing, in the Deathless shall abide.[06,158]

It has a subtlety hardly found in the Yoga Sutras. All-oneness, *kaivalya,* is handled more poetically in a Vedic hymn on the divine power than in the Sutras. It is portrayed as the sun:

In his bright chariot, set eternally to make All One
  A golden vehicle for peace, decked with sparkling light,
The Giver of Spiritual Life has come, resplendent as a Sun,
  Mantling the murky spheres below with his uplifted might.[06, 138]

And *Puruṣa,* seen as archetypal man, is portrayed in the Muṇḍaka Upanishad with much more of a universal flourish than in the Sutras:

Fire is his head, his eyes the sun and moon,
  the quarters are his ears,
  his speech is the revealed Word,
the wind his breath, his heart the universe,
and from his feet the archetypal earth.
He is in truth the essence of all beings.[06,148]

An explicit reference here to the four quarters or cycles of creative functions (Chapter Ten) is fairly frequent in the Upanishads, although not so explicit in the Yoga Sutras. Yet Whiteman emphasised that even abstruse points in the Upanishads are not theory or conjecture: "It is given as the fruit of direct perception and analysis, open to anyone not fixated in incompatible beliefs."[06,153] Clearly, Husserl's phenomenology was anticipated by millennia.

The generally unacknowledged influence of Gotama Buddha in the Yoga Sutras is a special feature of Whiteman's treatment. As noted in the previous chapter, soon after the Buddha's life there was a *rapprochement* between Buddhism and Brahmanism, which subsequently became lost during the period of the Christian/Sāṅkhyan Era, with the result that the Sutras fell into the orbit of Hinduism. English translations tend to be made exclusively within that orbit. In a chapter of *Universal Theology* (2006) Whiteman explored the recorded work and experience of Gotama Buddha, which had fed into the thinking of the authors of the Yoga Sutras at that time.[06,160] Yet he pointed out that several key words in the Buddhist Pali cannon have been subject to historical and conceptual distortion, not least through mistranslation of Pali words into Sanskrit[93,8], as happened in the writing of the Yoga Sutras. Some of these words need to be considered.

Atta, in Pali generally means "self", a "source of attaching identification"; the Sanskrit word *ātma,* breath, has a deeper sense, in the Upanishads generally meaning "the Divine Spirit or the same individualised, as a core identity or 'soul'."[06,162] Unlike *ātma, atta* implies a block to spiritual insight and progress on account of the tendency to attachment (regarded as a taint in the Yoga Sutras), hence the famous passage in the Buddhist *Dhammapada* translated by Whiteman as "All pure spiritual characteristics (*dhamma*) are selfless (*anatta*)."[06,162] In Chapter Eighteen, *dhamma* was rendered as "pure purposes or conceptualisings"; the word *dhamma* (*dharma* in Sanskrit) has a comprehensive meaning including right purpose, disposition, teaching or doctrine, most broadly "used to characterise the entire system of spiritual Truth and Wisdom taught by Gotama Buddha."[06,162] In all senses it transcends or is free of attaching selfhood and self-seeking, which must be thrown off.

Nevertheless, *atta* had a double meaning in the *Dhammapada,* translated by Whiteman as "self (*atta*) is indeed the Saviour of self. What other saviour can there be? By well taming self one gains a Saviour hard to gain." This points to the development of "a 'higher', unattached and wisely directing 'Self'."[06,164]

In apparent contradiction to the declaration, "all *dhamma* are *anatta*", are numerous declarations in Pali texts that "all *dukkha* (kinds of illfare) are *anatta.*"[06,164] Whiteman wrote that "the common rendering of *dukkha* as 'suffering' is damagingly wrong, suggesting that the relief of worldly suffering is the basis of Gotama Buddha's teaching... The *dukkha* which we must overcome if we are to proceed to an ultimate wisdom is simply whatever keeps us on an unwise but initially satisfying path. Suffering comes in when it is time for us to take a new stock of our 'self'."[06,163] Therefore the various kinds of *dukkha* arising from attachment are of selfhood in its attached state, but not of Self in its unattached, developed state.

Whiteman also discussed the frequent saying, "all *anicca* is *dukkha*"[06,164]. The rather obscure word *nicca* could be taken to mean "timeless", so that the physical body, for example, "is without 'eternal' character (*anicca*), subject to injury of many kinds, damage and destruction."[06,167] The saying could then be taken to mean that all transitoriness is a hindrance to spiritual progress.

In his translations and commentaries of the Vedas, Upanishads, Pali texts, and the subsequent Yoga Sutras, Whiteman kept emphasising the need for insightful interpretation of physical descriptions. The need is especially evident in the six Buddhist *abhiññā* or over-knowledges, a systematic treatment of skills that underlie the Yoga Sutras.[06,86] The first includes powers of passing through obstacles, walking over water, travelling through air; yet, following immediately on the analogy of pulling a reed from its sheath (Chapter Four), clearly this is a description of separative experience, not physical yogic magic.

In the second *abhiññā* mention is made of "the heavenly ear-function", a power that extends "far and near," which could be interpreted as telepathy. Likewise the third: in exploring one's heart and the heart of others "one perceives the other mind to be 'angry or peaceful, attentive or wandering, steadfast or wavering, spiritually released or not released'."[06,167] This may be seen as empathy, but perhaps intermingled with telepathy. The fourth *abhiññā* deals with retrocognition, personality structure and the disclosure of co-minds, a skill that detects coming-and-going in the house analogy of the fifth *abhiññā* (Chapter Twelve). Yet in all these parabolic analogies the focus is on undistracted spiritual development, so the sixth and last *abhiññā* exposes our "falling short of spirituality" through emotional attachment, failure of recollection, and ignorance in "thinking of our present life in this world as exhibiting our true and final identity, our core identity being hidden in a succession of personality-phases (*jātī,* 'births')".[06,87]

The need for insightful interpretation of physical descriptions is just as great when approaching Western mysticism. This is a central theme in the following chapters.

CHAPTER TWENTY-THREE

THE MINOAN CULTURE

Western mysticism could be said to have started with the Minoan civilisation, commencing around 2000 BCE. There are no decipherable writings from this Aegean civilisation, but it produced outstanding artefacts that reveal much of the life and thinking of the time. The Roman historian Diodoros wrote of the Cretan's claim that "divine honours, sacrifices and rites involved in the Mysteries were transmitted from Crete to other peoples," a passage that Whiteman found to be confirmed in their artfacts.[06,117] Even if we have no decipherable written accounts, "the pictograms in the Aegean civilisation are in fact more obviously descriptive of mystical teachings than a description using words with a customary this-world meaning is likely to be."[06,118]

A feature of the images is the depiction of four items: four pillars, four people or groups, four divisions. Whiteman interpreted this as representing the pervasive understanding of the cycle of four functions in creativity: purpose, choice of means, action, and incubation (Chapter Ten). Incubation may be followed by upward growth into an improved cycle, which may be represented by a central upstanding fifth figure in the pictograms. It could be represented by the thumb pointing above the four fingers indicating this generating potential; in later Minoan times Whiteman noted that "this five-fold pattern was duplicated and represented by the fingers and thumb of both hands (the *dactyls*), held up in a gesture of dedication."[00,47] A fifth upward element was also recognised in the Upanishads[06,119].

A fresco from the Palace of Knossos (Evans 1921-35) was depicted showing a temple with three shrines and four pillars.[00,119] Two central pillars are grouped together and elevated above the other two against an azure background. Whiteman assumed these to represent the first two functions, East and South, where the (divine) sun is in the sky. As discussed in Chapter Ten, he characterised these functions as Faith and Obedience, with male and female qualities interconnected, since purpose and means "cannot act otherwise than together. They constitute what is called the 'sacred marriage'."[06,120] (His retrocognitive experience as an apprentice Minoan priestess, cited in Chapter Fourteen, records pondering over the Minoan understanding of the sacred marriage.) In confirmation of the male and female interpretation, a rosette below the pillars contains small streaks or rays on one side and round dots on the other, emblems associated with male and female respectively in other pictograms.

Figure 23-1 Fresco from the Palace of Knossos. Evans.

The third and fourth functions, West and North, were considered by Whiteman to be represented by the black pillars in separate shrines to either side. The grouping of the first and second functions into the central shrine results in only three shrines, which Whiteman took to represent a three-level system of voluntary powers, Faith/obedience (E + S), Active Recollection (W), Continuous Recollection (N)[00,52]. These are the three foundational skills discussed in Chapter Sixteen.

The four functions and the fifth upward element are shown, Whiteman considered, in the accompanying pictogram from a signet ring in the collection of the National Museum of Denmark, Copenhagen (Nilsson 1927).[06,122] As is typical, four figures are shown which seem to represent the two male functions, East and West, and the two female functions South and North, with the fifth upward element represented by the Great Mother. Streaks are above the males, arcs of dots above the females. The females are standing with hands indicating giving, the males are kneeling with hands indicating receiving. The general import may be considered to be female, under the divinity of the Great Mother, being "born from the female fourth function, and stabilised by the female second function."[06,122] Behind the females seems to be fruitful vegetation;

accompanying the males are biconcave shields (the standard shape of Minoan shields), which Whiteman considered "can be reasonably taken to represent protection needed for the cycles in the upper and lower parts of the spiritual body."[06,122] The shape of the shields may suggest the form of the double cycle of *chakras* figured in Chapter Seventeen.

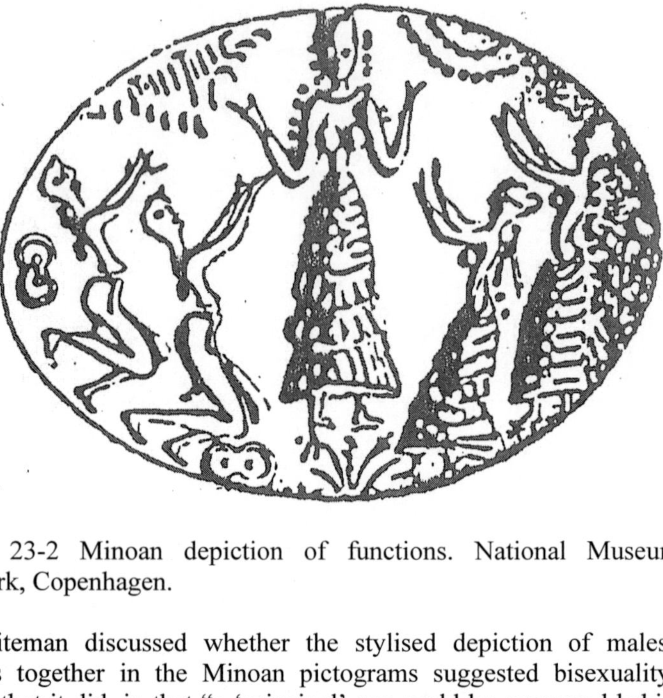

Figure 23-2 Minoan depiction of functions. National Museum of Denmark, Copenhagen.

Whiteman discussed whether the stylised depiction of males and females together in the Minoan pictograms suggested bisexuality. He argued that it did, in that "a 'principal' sex could be expressed habitually in the physical or spiritual body (as the case may be), and an 'auxiliary' sex would operate occasionally in the mind, spirit or physical body, as the case may be."[06,120] There seem to be hints of bisexuality in many forms, such as the frequent depiction of a double axe, biconcave in the centre to form the shape of an X. The outwardly turned cutting edges may stand for "male" (E and W), the inwardly turned arcs "female" (S and N).[06,12] Minoan art is striking in giving equality and complementarity to the sexes, if not female predominance. It seems to be the product of a uniquely enlightened civilisation.

Whiteman's chapter on the Minoans was concluded with the question, what is religion? He accepted the *Oxford English Dictionary* definition of religion as, "A particular system of faith and worship." Listing the comments of several standard authorities, he showed a general consensus that Minoan religion was a kind of nature worship geared to the seasonal cycle, but lacking any deeper system. The spiritual, mystical core of the religion seems to have been entirely missed:

> Their words (and use of the term "religion") hardly suggests anything out of, superior to, or hidden in this material world... the people of the time did indeed have a "system" of deeper insights beyond the everyday observation and manipulation of objects in this world. And that system, the character and details of which they were hoping to communicate parabolically, was surely concerned with the growth of spiritual perception and powers beyond those of normal everyday life in this physical world and body. Above all, they laid emphasis on the ideally "female" open and selfless response, givingness, and timeless contemplation of wholeness and values, from which the outgoing creativity of man is (or ought to be) born.[06,133]

Chapter Twenty-Four

The "People": Isaiah and the Psalms

The three previous chapters showed translators and authorities failing to see non-physical or mystical meanings behind supposedly physical descriptions in their texts. Not surprisingly, translators and commentators of the Old and New Testaments fall into the same mould. Whiteman selected passages from two books of the Old Testament to explore their overlooked mystical content, the Book of Isaiah (first part) and the Psalms. A mystical interpretation is possible especially when the passages are seen in the light of a corporate structure of personality rather than as affairs of a population of physical people.

As a result, his interpretation of commonly-used words is often radically different from what is usual, yet "this is chiefly because this-world meanings have customarily been chosen instead of parabolic ones as indicated by the verbal roots and required by the context."[06,200] In biblical Hebrew, "words very often have both concrete and abstract meanings. Literal meanings will then almost always be wrong."[06,185]

To ordinary reading of *The New English Bible* (1970), the second chapter of Isaiah starts with nations and peoples going for instruction to "the mountain of the Lord." Whiteman's interpretation "in line with what is known about mystical experience" is about elements of a personality being raised and guided to the "source of the Right and Good." That interpretation is continued for the rest of the chapter; the "people of Jacob" stands for "any individual who has entered the mystical way by 'supplanting' [the meaning of the name Jacob] the earlier this-worldly phase."[06,181] According to his analysis, "we can see that Isaiah is mapping out the stages of spiritual development (up to the mystical) in precisely the same way as in the reports of various Christian mystics, and closely corresponding also to the stages mapped out by Gotama Buddha."[06,182]

Warnings to Jerusalem and Judah in the third chapter were taken to be symbolic (the place-names mean peaceful and praiseworthy), for "the concern of this chapter is not with historical events, but with the loss of security, through an upset of values and personal control, afflicting the prospective mystic prior to disclosure of the divine Source."[06,182] Afflictions of this kind were discussed in Chapter Sixteen as ultimate contests; extracts from Isaiah on fateful "the end of days" are given in Chapter Twenty-Nine.

The "blistering attack" on women's fashions that follows is seen as "a later editorial addition, certain writings of some scandalised non-mystic being thought appropriate here. For the malice is not in those who conform to fashion and wish to become if possible more beautiful, but in the critic."[06,183] Coming from a mystic, this comment in favour of fashion-dressing is interesting. Evidently the thought behind it is that self-beautification can be objective *dhamma* that is *anatta* (Chapters Seventeen and Twenty); a pure uplifting and aesthetic purpose in dressing can transcend selfhood (a central theme of my novel, *Brief Authority,* 2012.)

Whiteman held that in the sixth chapter it becomes explicit that Isaiah is recording mystical or transcendental experience, with a disclosure of "the Lord". He is told to go and tell "this people" that they may listen but will not understand, look but never know; they cannot see. And they will continue this way to utter ruin and desolation. The meaning Whiteman places on this puzzling passage is that "this people" refers to elements of the personality (co-minds) that are unable to comprehend the disclosure of the divine Source; their negative influence must be faced, being "concerned with outmoded motivations, interpretations and habits." The individual's "higher part", in taking command, "must learn to overcome the 'lower' part permanently."[06,187] So in due course, "the unconformable 'people' will have been in effect removed, or set apart, leaving the core identity [Chapter Twelve] like the trunk of a tree stripped of its leaves."[06,188] Isaiah's visions of these upheavals are discussed in Chapter Twenty-nine.

Whiteman saw such passages as an allegory of the corporate structure of personality. Just as in the ancient word a king was held to guard his people and be a mediator between God and them, so "the released core identity of any individual, or the ruling identity acting for the unreleased personality of the individual, was called a 'king', and the contributory minds were the 'people' (or 'peopling')."[06,98] This throws revolutionary light on the famous declaration in the Isaiah ninth chapter (as in the Authorised King James Version): "The people that walked in darkness have seen a great light ... For unto us a child is born, unto us a son is given ..." Whiteman interpreted this as the mystic's personality transformed from walking in spiritual darkness to seeing the great light of Divinity through the rare disclosure of a core identity, the "son".[06,189] This, one's "ultimate mystical identity", is carried from life to life, either archetypally male

(Son) or female (Daughter). In Isaiah's account a "son" is given in the reformed life, and the "government" of the personality is "on his shoulder". This idea is carried through to Whiteman's treatment of the New Testament (next chapter), "Jesus" standing as the Son in the "government" of the personality.

In the tenth chapter of Isaiah, verses 27 to 32 list twelve towns through which the Assyrian army is said to advance. Whiteman pointed out that there is play on the word *ashūr,* which can mean either Assyria or "the way, manner of life, or conduct."[06,190] He found "an astonishingly detailed and accurate account of the 12-fold creative cycle, represented as a mythical advance of the Assyrian army through twelve towns (with significant names) towards Jerusalem ('foundation of peace'), and ending with a salute."[06,190] Its inner meaning describes an advance of the personality towards mystical gain. A simplified table shows this, based on the table in Chapter Ten of functions (quarters), subfunctions and psychological terms, added to which are the meanings of the names of the towns. The number pairs denote complementary essence and existence.

E 16 purpose actuation: enthusiasm – "comes to"
E 15 purpose becoming: inspiration – "passes through"
E 14 purpose grounding: latent energy – "lays up baggage"
S 26 means actuation: feeling-response, Obedience – "there is lodging for us"
S 25 means becoming: orderly shaping – "is concerned, God-fearing"
S 24 means grounding: judgement and choice – "is put to the test"
W 46 grounding actuation: extroversion – "raise your voice, unfold"
W 45 grounding becoming: application – "be attentive"
W – outward identification – "bow down, respond"
N 36 fulfilment actuation: non-attachment – "is flying away"
N 35 fulfilment becoming: time-stopping, Active Recollection – "make strong and secure"
N 34 fulfilment grounding: expansion, Continuous Recollection – "stand firm, continually now"

The eleventh chapter opens with a description of the fourfold creative cycle. There is settled upon the Son "the spirit of the Lord (E), the spirit of dexterity and wisdom (S), the spirit of design and power (W), and the spirit of reflection and reverent submission to [the guidance of] the Lord (N)."[06,193] (The Hebrew words in this passage are here omitted.) The chapter continues with "some further subtle mystical teachings concerning the life of a mystic in this world patterned on 'the Son'." It includes the famous passage of the wolf dwelling with the lamb (Hebrew words are omitted): "His delight is in a reverent submission to the Lord... So [within the personality of a 'son'] the wolf will [come to] dwell with the lamb ... while a little child is guiding them."[06,194]

Whiteman saw significance in the reference to a child, discussed in Chapter Seventeen. "Reports of mystics show that when a mystic is transformed as 'Son' or 'Daughter' it is not at all as a copy of the physical personality in appearance, but is usually of some age between about seven and fourteen, and may go back to shortly after birth. But the conviction will be that such bodily form truly represents the lasting *core identity*."[06,194] As noted in Chapter Seventeen, he reported being characteristically transformed as a girl of that age, being mystically a "Daughter", noting that "a spiritual 'body' will conform to a certain basic state of mind at all times, not the reverse."[06,72]

A translation of four psalms continues analysing the figurative meaning of place and people. For the Psalmist as well as for Isaiah, "the 'people' are those influences in the personality that are normally identified by the individual as his/her own (rightly or wrongly). On the other hand, those kinds of influence which are at first from 'outside', and may or may not be eventually accepted for the core identity, are called 'foreigners'."[06,179] Whiteman's translation of the opening of Psalm 2 (dissimilar from both *The New English Bible* and the contrasting Authorised Version) is: "Why do 'foreigners' within us get emotional, And our 'people's' thoughts are often out of joint?"[06,201] Verses 7 and 12 call for more order through the Son: "The Lord has said to me: 'Thou art my Son' ... Give love and deference to the heavenly Son."[06,201]

Whiteman interpreted the whole psalm as declaring that behind disruptive forces in the personality, with a succession of inner contests, there is divine purpose. Control or reformation is brought about by patterning on an archetypal Son; obstruction to unity falls away with love and deference. Even when the mystic has reached the stage of unitive transformation, he/she is still open at times to being undermined. So, in translation, Psalm 110 starts:

> This is the Lord's instruction: "Honoured Sir,
> At my right hand take your seat
> And I will make your adversaries
> A lasting support for your feet."[06,203]

The "people" (making up the individual personality) are to conform without duress, with a breaking down of worldly "kings", "presumably sources with a strong tendency to be ruled by worldly success or failure, pleasure or pain."[06,205] So the mystic, "keeping in mind the Right and

Good [being at the Lord's right hand] will be victorious through refreshment as from a 'river' of pure spirit."[06,203]

Psalm 45 addresses a "King" which Whiteman initially interprets as "the archetypal Saviour-King", not a physical king; perhaps it has a similar meaning to *Puruṣa.* But in verse 7 he believed the meaning changes to "that of a mystic revealed as being in the state of Transforming Union as a Son, with contributory minds."[06,211]

> So therefore God has chosen to anoint thee
> With joy, above those in thy conjoined throng.[06,210]

The psalm then turns to a daughter, which Whiteman interprets as the fully recognised female auxiliary sex of the Son (Chapter Seventeen). The text reads that a daughter "has all kinds of wealth within."[06,213] Whiteman interprets this as the feminine second and fourth functions complementing and making effective the male first and third functions. The psalm, which at first sight "might seem no more than a royal marriage ode"[06,209] can then be interpreted as an ode to the sacred marriage (Chapter Twenty-two).

CHAPTER TWENTY-FIVE

THE "SON": ST PAUL AND ST JOHN

The beginning of Chapter Twenty-four discussed Whiteman's view that translations of scriptures "have almost invariably been made by scholars lacking mystical experience."[06,viii] His attempts to correct shortcomings led to familiar words and passages gaining unfamiliar meanings, as in apocalyptic passages which he considered were descriptions of upheaval in ultimate contests. These contests were discussed in Chapter Sixteen. He took as an example St Matthew's "Little Apocalypse" (Matt. 24:3-44). The prophesy that "This generation shall not pass, till all these things be fulfilled," did not turn out to be physically fulfilled. Whiteman interpreted the passage as describing an ultimate contest:

> (I)t plainly describes a period of chaos before the disclosure of the Divine Source. It begins with a "horror of desolation," when one is advised to "flee into the mountains" (put one's trust in the higher values?). Startling phenomena ("great signs and wonders") could deceive even those destined to come through the ordeal; and "tribulations" will be greater than ever experienced before. But no mistake can be made regarding "the appearing of the Son of Man in heaven"; for it is like the dawn, spreading light from the east to the west. Everyone should "watch therefore" and "be ready"; for "the Son of Man comes at an hour you do not expect."[68a,646]

His attempts at textual corrections are especially notable in his long chapter on the mystical teachings of St Paul. The meaning of "sin" (*hamartia*) and "repentance" (*metanoia*) in Paul's epistles was discussed in Chapter Nineteen. Another word, *pistis,* is customarily rendered in English as "faith". Whiteman traced the word back to the verb *peitho,* "to win over", so "*pistis* suggests 'being won over' or 'obeying' (what is right and good)." He saw this as standing for "the necessary attitude of mind or way of responding to the problems of life if one is to advance to the ultimately True in belief and ultimately Good in valuation—in other words, to *living* the 'mystical' (as understood today)."[06,215]

More radical is his translation of *Christos Iēsous* as Anointed Saviour, not the customary Christ Jesus. He wrote:

> The word *Iēsous* was chosen by the Septuagint translators of the *Books of Moses* as standing for Joshua (sometimes written *Yēshua*‘ in Hebrew, meaning “help of the Lord”). Moreover, the ending –ous characterises the genitive case, masculine singular, so the meaning suggested is “He of (i.e. son of) Ieso”, where Ieso would be the later (Ionic) form of an earlier Iaso, goddess of healing. And *Christos* is merely the Greek word for “anointed”, that is, dedicated for some divinely appointed task. When Paul spoke of Iesous Christos, therefore, this is what he and his hearers would have thought of, not the Gospel stories current a century later. To avoid any such anachronism, the name will be given here as the “Anointed Saviour”.[06,216]

In Whiteman’s view, Paul evidently saw spiritual powers to belong to a “universal Humanity” (presumably equivalent to a meaning of *Puruṣa* in Chapter Twenty-one), latent at “the core of our ultimate ‘real being’ and in the inspiration for our life in the Right and Good. The name *Iēsous* can be said to focus on the ‘higher life’ and upliftment to it, the name *Christos* on the dedication, and the title Lord on the perfection and power.”[06,218]

His seemingly dismissive phrase “Gospel stories” needs explaining. He found “intractable difficulties” in seeing the synoptic gospels (Matthew, Mark, Luke) “as basically historical narratives concerning the life of a physically-living human individual.”[06,xi] He presented arguments common to radical Higher Criticism that key features of the synoptic gospels have a basis in ancient mythology, finding support in Mark 4:11 that “for those without (understanding of the mysteries) all is presented in parables.”[06.xii; 06,63] He maintained that “To anyone familiar with Greek,” the words *Iēsous Christos* “would surely suggest something universal rather than an individual man in this world. But it would be a useful title for the wider public, who could be attracted by parabolic stories (about an individual man), whereas the more abstract teachings could repel many.”[06,76]

Nevertheless, Whiteman considered that if the synoptic gospels were to be understood as “demanding that a certain historical individual named Jesus be identified in a unique, total and eternal way with the universal deity,” then they “would rule out the possibility of a universal theology, covering all major religions of ancient times.”[06,x]

For reasons discussed later, he regarded the accounts of a historical Jesus in John’s gospel also to be interpolations and parabolic. There is no clear mention of an execution of a man Jesus in the letters of Paul, or, as Whiteman pointed out, in the letters of Peter, James or John. He considered the standard idea of a “crucifixion” to have no discernible historical basis, but to have been developed from long-established symbolism widespread in the region: the Tree of Life. Ideas such as a "Lord of the faithful tree" being slain together with a malefactor "must have permeated the near East for millennia, since in the first letter of Peter (Ch. 2,v.24) we read that the *Christos* 'bore out faultiness [*hamartia*] in his body on a tree ... that we might die to faultiness and live to rightfulness'. So when Paul spoke of the *Christos* as being 'crucified', his hearers would most probably have known at once to what he was referring."[06,225]

Reference to a "cross" in the epistles, Whiteman believed, "can reasonably be understood as standing for the fourfold cycle in the conquest of suffering... it is surely clear that Paul is referring to the archetypally divine (but also human) power behind the fourfold 'saving' cycle, by means of which we are raised through suffering."[06,77] A sacrificial victim is "the archetype of regeneration."[00,30] A passage from 1 Corinthians 11:12 was thought by Whiteman to be "even more obviously indicative of the fourfold creative cycle: 'In the Lord, neither is woman without man nor man without woman. For as woman is [brought] forth from man, so man is by woman; but all things are of God'."[06,224] The complementary east-west male and north-south female components of the cycle have been discussed in several preceding chapters.

In the Acts of the Apostles 26:9 Paul is reported to have used the name Jesus of Nazareth; Whiteman pointed out that the name Nazareth is evidently derived from the Hebrew *nāzūr,* "to consecrate oneself", equivalent of the Greek *Christos*[06,xiii]. As Jesus was said to have been born in Bethlehem, city of David, the appellation "of Nazareth" might have the same inner meaning of the word "Christ". Whiteman's view was that the "anointed saviour" Paul wrote about was not so much a person, "Jesus", as standing for the released "Son" as core individuality in relation to the Source of the Right and Good as "Father". This would place Paul as a mystic who, on the road to Damascus had experienced the vision of archetypal light, characteristic of mystical initiation (Chapter Twenty-seven). Whiteman translated a famous passage of Paul in Galatians 1:15-16 as, "When He who had set me apart ... was pleased to reveal his Son in me." He noted that "the translation 'to me', sometimes found, has no warrant whatever in the original Greek, and seems to result from the translator being unable to make sense of the passage except on the basis of one-level naturalism."[86a,142]

It was in this mystical light that Whiteman viewed Paul's letters; yet Swedenborg saw it differently. He regarded Paul "as little short of demonic"[06,xiv] because he made no reference to the synoptic gospels.

Evidently Swedenborg did not realise that it would have been historically impossible for Paul to have done so, the gospels having been written after the epistles. Neither did he question the historical truth of the gospels, although this was at a time before Higher Criticism had taken effect.

Whiteman held that another Greek word mistranslated with damaging consequences is *parousa,* conventionally rendered in English as "coming" or "second Coming". "Coming" follows the practice of application and non-attachment (as in the Yoga Sutras), leading to release, transformation, rebirth. "The actual *transformation,* when it comes, is called a *parousa.*"[06,218] His translation of 2 Thessalonians 2 is then:

> Now concerning the transformation (*parousa*) and our conformation to Him [the Anointed Saviour-Lord], we beg you, bretheren, not to jump to conclusions that the "day of the Lord" is come. That day will not come unless a non-attachment comes first, and the materialistic (*anomous*) man is revealed as destined for destruction. Such a man opposes, and exalts himself against all that is named "God" or an object for worship.[06,218]

Whiteman's rendering of *anomous* as "materialistic" comes from *nomē* meaning "a feeding". So, "Anyone who is habitually not guided by the right and good is then not properly 'fed' and so can be said to manifest *anomia.* A materialist today is such because he takes dead matter and natural law, along with the motives of self, to be the ultimate realities."[06,219]

The "ultimate realities" for Paul were settled, as they are for any developed mystic, by his direct experience, as in Whiteman's translation of 2 Corinthians 1:5: "When the trials of the Anointed One (*Christos*) abound in us, then, through the Anointed One, we share also abundantly in the spiritual support given."[06,91] The same was recorded in the letters of St John, to which Whiteman devoted another chapter.

Uncertainty exists about the identity/ies of the disciple John, the author of the Gospel of St John, the author of letters of "John the Elder", and the author of The Revelation of St John the Divine. Based on textual analysis, Whiteman considered that, barring considerable interpolations, the author of the gospel and the letters was one and the same person, and one who had been visited by Paul about 49 CE.

Whiteman saw the same errors in conventional translation and understanding in John's writing as he found in Paul's. In the famous "test the spirits whether they are of God" passage in John's first letter (4:1-3), his translation is: "Every spirit which conforms to the idea that an Anointed Saviour has been settled in the flesh is of God; and every spirit which does not accept [a universal] Saviour (*Iēsous*) is not of God." He found difficulty with the word "flesh", *sarx.* The word would be "grievously misunderstood" if taken to apply to some individual's physical body, namely a historical "Jesus". This would be "a denial of 'spirit', as in materialism."[06,239]

Much of John's gospel must have been added after his death, Whiteman argued. For example, the bitter anti-Jewish passages seem likely to have come from a time of antagonism between Jews and Christians around the time of the Bar-Cochba rebellion against the Romans, 132-135 CE, when Christians tended to side with Rome. This happened after John's death. The name of the fabled betrayer of Jesus, Judas Iscariot, is derived from the Greek name for Jews (*Yihūdī*), and *ish/kārōt* means "man of destruction".[06,241] Whiteman pointed out that this "could hardly have been a baptismal name; but it describes his function in the anti-Jew story." He asked, "Could John, a Jew, have written in that spirit?"[06,254]

From these manipulations of Christian text, it is understandable that Whiteman's study of ancient mystical religions stopped with early Christianity, after which he saw religious thinking corrupted by meddling, dogma and theorising (Chapter One). Nevertheless, ignoring the insertions, he believed that what remains can be seen as John's account of "his transforming experience (or experiences) of the Divine Source or Godhead." This is reflected in the opening of his gospel, translated by Whiteman as: "Archetypally (*en archē*) there was Reason (*logos*). Reason was with God, and Reason was God ... And Reason passed into 'embodied humanity' (*sarx*) and was staged (*eskēnōsen*) in us [as a divinity sensed within]."[06,241] Whiteman treats the remaining parts of the gospel that he considered authentic as accounts of a mystic to whom the Divine Source has been disclosed. Included in the disclosure is "the mystic's core identity as that of either a 'Son' or a 'Daughter'."[06,252] This was discussed in Chapters Seventeen and Twenty-four; the meaning of a "Son" in a transcendent sense was evidently long-established in Hebrew religious thinking.

In John's gospel, Whiteman saw *Iēsous* as standing for the inner Son, and the "disciples" standing for the compound personality including co-minds. The "Father" stands for the Source of Right and Good, and "as mystically conceived, is referred to not only in Christian writings, but also in the Old Testament (e.g. Isaiah 9:6) and the Upanishads."[06,xxxiii] He could be understood as the unseen Source behind the motivations of the Son, and "must be thought of as 'male' because the emanation is creative (from a W above)."[06,xxxiii] In some passages it is as if Jesus (the Son) is speaking to the Father, for example, "They [disciples] are not of the world, even as I am not of the world. But as thou hast sent me into the world, so have I sent them. I do not concern myself for these only, but also for those who believe with reason in me. So all may be One, as thou, Father art in me, and I in thee."[06,253] The "One" may perhaps recall *kaivalya* in Chapter Twenty-one.

Evidence for John's thinking in terms of the four-fold cycle is found, Whiteman maintained, in passages such as in the fourteenth chapter when the "Son" speaks as "Jesus": "I am the way forward (E), with full comprehension (S), and leading to life (W). He who has seen me [as such a 'Son'] has beheld the Father [at a higher level of causation]."[06,250]

Whiteman concluded that "the exceptional value of John's Gospel (the authentic chapters) is in his account of the Mystical Way and the Divine Humanity, even extending that of St Paul."[06,254] Looking back on the chapters in this Part Three, it seems that a common set of ideas or metaphors was current in the ancient world, which became misunderstood, misinterpreted, corrupted, or lost to materialism at the start of what is called the Christian Era. Perhaps it might just as well be called the Sāṅkhyan Era, considering what was discussed in Chapter Twenty-one. This trend now seems to be in reverse, indicated by the gradual renewed interest in authentic mystical understanding (e.g. Kelly, 2007; Marshall, 2015). Combined with maturation in scientific thinking, in the form of post-materialist science (Beauregard *et al.*, 2014), the times seem ready for the emergence of a new science of mysticism and theoretically secured psychical research.

PART IV

ELEMENTS OF MYSTICAL EXPERIENCE

CHAPTER TWENTY-SIX

LIFE IN OTHER WORLDS

"Complex in the extreme"[06,276] was how Whiteman reported meetings with people in other worlds. This could be said of the whole subject of "life in the other worlds", the subtitle of his last book.[06] Spread through most of his books are records of otherworldly experiences, sometimes accompanied by a "stringent warning" against their being read as if describing physical scenes and experiences. As most people's experience does not encompass otherworld states, there is a particular need to read the reports with a clear and cleared mind to grasp them rightly.

There is an immediate challenge to understand reports that describe Whiteman's separative experience in the form of a girl. His observation was noted in Chapters Seventeen and Twenty-four that "when a mystic is transformed as 'Son' or 'Daughter' it is not at all as a copy of the physical personality in appearance, but is usually of some age between about seven and fourteen."[06,194] Being mystically a "Daughter" and characteristically transformed as a girl of that age, the point to emphasise is that "life in the other worlds" can be unrecognisably different from life in the physical world.

This requires appreciating that "while in this world we think of our body as causing what happens in the mind, and thus carrying our individuality from one moment to another, the reverse is the case when we have become settled in a non-physical world. For there, the body in which we find ourselves is obviously determined conjointly by our settled motivational character or needs at the time, and the surroundings (spiritual level, and particular scene)."[06,35] A non-physical body might not even be discerned. Whiteman emphasised that "*we will not be able* to discern even our own limbs (that is, the limbs of our proper separated body) unless our mind is in some way orientated according to the spiritual significance and archetypal localization of 'limbs'. Much the same will apply to objects around, such as 'trees' and 'water'."[61,16]

Whiteman's other-world accounts generally differ markedly from the numerous out-of-body reports in the literature, which tend to be physical-like (Chapter Five). Notable are differences from the reports by Swedenborg. Whiteman's central quest was for gaining spiritually transformative upliftment and release from the physical, while Swedenborg could be described in this context as an investigative reporter, hardly concerned with his own transformation. Whiteman viewed life in "the other worlds" to be heavily dependent on the aims and state of the observer, very much more than it is in the physical world. He wrote, "The difference between one 'world' and another ... depends upon the state of life in which the observer happens to be settled."[06,3]

For most people, settled only in physical life, the "world-states" usually encountered in separation are classifiable as "duplicate", discussed mainly in Chapter Seven. Here items in the physical world seem to be copied. It is not too difficult to get a sense from these reports of what is being experienced, but a "higher world" experience takes ordinary understanding and allegorical interpretation about as far as it can go. Swedenborg's accounts do not go this far, perhaps deliberately if he anticipated readers' incredulity.

Rather than "other-world" experience, many of Whiteman's accounts could be described even as ante-world experience, as in his account of a vision of archetypal light.

> The separation did not begin from a dream state, but during the night I became fully awake to the quality of [a continuous state of] directed Obedience, now stronger and more clearly known. Almost immediately, the separated form was drawn upwards, quickly, and as if through a great distance, but without any loss of steady balance or of the feeling of peace.
>
> All at once, without any further change, my eyes were opened. Above and in front, yet in me, of me, and around, was the Glory of the Archetypal Light. Nothing can be more truly Light, since that Light makes all other light to be light; nor is it flat material light, but a creative light of Life itself, streaming forth in Love and Understanding, and forming all other lives out of its substance: a Light become Life not through addition to material light, but by the removal of the impurity of fixation.
>
> What can we say of the contemplating soul? Formed by the Life itself, she was yet face to face with its Glory, poised and held aloft as in the air, while the Power streamed forth and past, shaping her form as to the wind—poised through the adoration of love and total submission, held aloft through His Majesty and Mercy...
>
> How can the Source be described? How its direction? Though upwards and forwards, it was not a geometrical direction, more or less so, and in relation to something else; but an absolute direction, exactly so, by its own archetypal nature. Source it was, of Life and Truth, being the source of all Ideas of Life and Truth, yet manifested as in space.[86a,146]

In Whiteman's writing on mystical experience, nothing is intended to be taken as remotely physical in this kind of description. Even directions are allegorical: "upwards" indicates upliftment, "forwards" indicates progression. Therefore one should not take the "contemplating soul" to indicate a person as normally understood, but rather as essence or core identity underlying inner life, with no manifestation at a physical level except indirectly in the expression of attitude and motivation.

It may be noted that as description of the Light experience progresses, mention of "I" and "me" disappears. In similar experiences, Whiteman notes that these terms are avoided since they could evoke "ideas of personality instead of concepts of uncorrupted core identity such as we all have deep within us."[06,16] This resonates with Gotama Buddha's teaching (Chapter Nineteen): "All pure spiritual characteristics (*dhamma*) are selfless (*anatta*)."[06,162] Core identity is a selfless state in terms of personality.

Other-world experience reported by Whiteman covers an enormous range, from ante-physical witnessing of archetypal light to the duplicate-physical meeting with Nancy, described in Chapter Five. Yet Whiteman found Nancy's apparent recognition of him to raise a problem. "In conformity with the mystical transformation by which, through orientation to the Source, all conditioning by the physical body is transcended, the appearance of my separated body in a stabilised separative state has been, after 1930, always so different from that of the physical body as to be unrecognisable..."[86a,73] Any tendency towards physical appearance in separation "evokes such distaste as to cause a quick lapse into a fantasy or waking state. The 'transformation' and consequent release from physical conditioning is maintained even in a duplicate state."[86a,74]

On the other hand he did write that at an intermediate level "one may take on the appearance of one's physical body so as to be recognised by other beings."[06,108] Presumably this happened in the Nancy meeting; it was also reported to occur in "thought separations", which are secondary separations where "I am aware of standing in a non-physical scene, in what is very like my normal physical personality, though no bodily details are focused on. That is to say, I am not aware of being otherwise than in normal bodily form."[06,264]

In the range of experiences from duplicate physical to transformative, Whiteman described manifestation spanning "lower worlds" to "higher worlds", as well as in the "intermediate level" just noted. This is a large subject that requires a chapter on its own.

CHAPTER TWENTY-SEVEN

LOWER AND HIGHER WORLDS

Lower and higher worlds were recognised as far back as the Rig Veda. Each world was typically regarded as divisible into three levels. A Vedic translation by Whiteman reads: "The three heavens are placed in him (Varuṇā, the Omnipresent) and three earths below."[06,106] The three heavens "are manifested to us as stages of spiritual development,"[06,306] corresponding to "(1) the overcoming of attachments by Active Recollection, (2) gaining the power to expand into wholes without losing sight and control of the contributory parts, in Continuous Recollection, and (3) an Ultimate Release, in which the Divine source, core identity and mystical transformation are revealed in the third heaven (or stage of development)."[06,288]

Three levels in the "lower worlds" were seen to correspond with "the three kinds of mental disorder, (W) conduct disorder, e.g. losing one's temper, selfishness, (N) neurosis, confusion escaped by defence-mechanisms, (E) and (S) psychosis, further distinguishable as paranoid or schizoid."[06,xxxv] Whiteman cited Isaiah 27: "In that day the Lord, with his terrible great and strong sword, will deal with leviathan the fugitive serpent, and with leviathan the tortuous serpent; and he will slay the dragon that is in the sea."[06,107] He considered the fugitive serpent to stand for conduct disorder, tortuous serpent for neurosis, and the dragon and sea as paranoia and schizophrenia. He commented, "Mystically, these disorders would be accounted for as the result of some kind of *possession,* i.e. relatively undeveloped CMs will have become dominant in the personality, without the IC having the power to objectify the presences and so throw them off."[06,197]

Isaiah's references to the "day of the Lord" or the "end of days", which occur in several other places in the Old Testament, were taken by Whiteman to stand for personality-remodelling, ultimate contests discussed in Chapter Sixteen. CMs are "being brought to light (dissociated and objectified), and so made to leave their previous close association with the individual in charge. Ones with a new outlook will take their place subcounsiously."[06,109]

He emphasised the need to distinguish between how the inhabitants of the lower worlds see themselves or their surroundings, and "how they appear in divine light when that is providentially given. In divine light an inhabitant may appear fiery, corpselike or leprous, according to deficiencies in his/her settled state of mind."[06,xxxv] Yet such divine disclosures, including 'ruinous or otherwise depressing surroundings ... would not be exhibited for those to whom such attitudes of mind and kinds of life were natural."[06,300] Swedenborg (Ryder, 2010) reported the same thing: those in a hellish state generally seemed oblivious of their situation and quite content to be where they are.

The creative functions operate differently in the different worlds, Whiteman noted. For example, "whereas, below, the quarters E and S (paranoid, schizoid) are markedly different in certain ways, in the higher spheres, E and S are as if married."[06,306] East and South, as discussed in Chapter Ten, were characterised as Faith and Obedience in the higher cycle, acting together in the "sacred marriage".

Here is an example of a reported experience in the lower world, seen in "divine light" after having passed through a higher world. As noted in Chapter Twenty-five, allegorical reading of these experiences is needed. Direction has inner significance: turning right and looking up are associated with high quality; moving down, left direction and turns are associated with the opposite.

> A long separation began with the clear sight of clouds in a bright sky. Recognition of the character of the state and reflection on it led to memory of the Source. Turning right and looking up, I saw the inner Sun, very bright, and significantly pouring forth golden light of divine quality. Looking along the level again, Obedience being more clearly awakened within, I found a scene of extraordinary character presented in front, slightly down and slightly to the left; the exact placing was felt to be highly significant of deeper spiritual realities. A large dingy building like a disused factory stood as the lower level, as in an excavated area, surrounded by yards and smaller sheds or outbuildings, all rather squalid in appearance. I recognised the "infernal" quality revealed in what was presented, but there was no fantasy or obsessive power in it, because the Source had not been forgotten, and the influences had been objectified in a divine light of dread, exposing the real nature and causes behind.
>
> In response to Obedience (the Source being still present with me, as it were) I moved forward and down, descending a slope covered with stones. Making my way among the buildings and sheds, I tried to find someone living. Eventually, in one of the sheds, I found a being like a man, but obscure, only a vague shape being discernible, though his presence was objective and unmistakable... A sexual connotation was obvious, as was also the danger and total unacceptability of intention emanating from him.

> The secure grounding in my proper being caused an instant and effortless dismissal of the persuasion, so that he no longer seemed to have such intention... His appearance was now shown in a more real light, his face in particular being clearly seen. It had apparently been attacked by some disease like leprosy, producing disfiguring ulcers. I moved back, away from him and towards the right, not panicking, but with a kind of calm horror at spiritual ugliness. For such an appearance, I could not help believing, is representative of confirmed disease of character. The experience then ended rather abruptly, as if the teaching I was to receive was now complete.[06,303]

In contrast is an example of awareness in the "higher spheres", which necessarily involved complete transformation in core identity. Again, allegorical reading is needed.

> During the evening I was standing quietly in the bedroom, in a settled state of Continuous Recollection and Obedience, when suddenly I became established in a mystical tertiary state while the physical world was almost entirely blotted out. I knew myself then in my spiritual body of love, standing similarly and without apparently having shifted my location, but in a boundless world of intelligible space and substance. Around me were other youthful female beings in whose life I participated, and great warmth of loving communion went out to them vividly from my heart and through my breasts, by virtue of their universal "giving" character and power. Reciprocally the inflow of life from those loving beings around gave a supreme comfort in knowing that I was now "home", where I really belonged, among loved and loving ones like myself.[06,310]

Other experiences, Whiteman noted, may be interpreted in the light of the Biblical "last times" and "day of judgement", or as contests in general. Fire is commonly encountered. An experience in primary separation is recounted as: "I came to what looked like the suburb of a big town. It was suggested within that something quite near was 'on fire'. Crowds of people were streaming along. The grass verges and houses were clear, and the faces of the people, mostly going the opposite way to me, could be fairly easily studied. They seemed under some pressure of intention to get somewhere."[06,287] He emphasised that the fire "must not thought of as in any way physical," and he interpreted the people as "being brought to the point where they could be prepared for the 'first heaven'."[06,288]

Whiteman recorded several experiences of seeing and touching people and pets who had "passed over", but it is not often clear what the total situation was. In two cases of seeing people who had died violently he stated that the experiences "were not of apparitions. The beings seen were not inserted, so to speak, in a normal scene around me. I was in fact somehow in the same 'world' with them, yet at the same time fully able to switch my mind at any time to the physical world still around me. In other words, the state was a variety of Tertiary Separation."[06,281]

In contrast, he recorded meeting individuals in a separated state who had been inwardly named as colleagues still physically living. Their appearance did not resemble their physical appearance and he surmised that he had met a contributory mind of the personality. "On the other hand," he wrote, "in meetings with individuals shortly after their physical death, the physically-living observer usually (but not always) attaches the name so as to agree with their appearance, not their secret trends of character or function in life." He commented, "It all depends on the circumstances."[06,285]

As observed at the beginning of the previous chapter, the topic of life in other worlds is "complex in the extreme".

Chapter Twenty-Eight

The Archetypal

The archetypal reaches into the core of Whiteman's thinking. "It must not be supposed," he emphasised, "that 'archetypal ideas' are a matter of 'thought' or 'imagination', and so 'unreal'." The archetypal is "*precisely what constitutes indubitable meaning.*"[06,93] This relates to a passage that was abbreviated in Chapters One and Nineteen, that "the word 'God' (*Theos*) must be taken to stand for the Archetypal Reason in all, including the 'I AM' and consciousness, if the ancient testimonies are to be understood." This was taken to be "a first great truth that must be admitted if one is to talk or write reasonably of Theology."[06,64]

The term I AM appears mainly in Whiteman's late work. In a Christian context the term is associated with God, or Jesus equated with God; but in Whiteman's usage it is applied to an individual's core identity or line of consciousness. He wrote,

> we have a *core identity* which carries (so to speak) the concept of "I am" (not "me"). As soon as we think of "me", our mind is focused on an object, which is *not* our core identity. We call it "my self"; and in general, when we adopt such an abiding focus on "self" (instead of complete openness and freedom from attachment) such "self" will somehow limit or even distort the characteristics of "I am"—as "consciousness" becomes "self-consciousness", "will" becomes "self-will", "centred" becomes "self-centred", and so on with almost every human characteristic."[06,269]

His identification of God-like "I AM" with an individual core identity may seem sacrilegious to a Christian, but this view could be held to underestimate the divine nature of core identity as Whiteman saw it. He wrote, "Though the term 'self' points, in the mystical view, to a counterfeiting of our 'real being,' one may accept that in every counterfeit there is hidden a more or less unacknowledged germ of the real, which in this case can be taken to be the I AM of the Godhead, individualised somehow by the divine potentiality."[00,145] Yet he thought it allowable to use first-person pronouns in this context "on the strict understanding that the *selfless real being,* patterned on the Divine Obedience and Faith, is being referred to and not anything patterned on or limited by things in the physical life."[00,146]

Evidently this is another example of all pure spiritual characteristics (*dhamma*) being selfless (*anatta*), as discussed in Chapter Nineteen and later. Core identity is selfless and archetypal, yet in Whiteman's usage it is nevertheless "established with a unique I AM."[06,75] It was seen as individual, and, as cited in Chapter Thirteen, "The idea of a physical death as an end to the individual's I AM seems to rest largely on ignorance or rejection of the evidence for 'other worlds' and the processes of separation (from the physical body), as well as any logical purpose in regard to the continuing life of individuals".[06,75]

Despite its archetypal nature, Whiteman observed that "such core identity, established with a unique I AM, is at first inevitably involved in attachments and identifications (by the third creative function, W). And only when a certain non-attachment has been practised, will a holistic release be granted and something of the character of the core identity which is being built up will come through clearly."[06,76] If the core identity is to have a developed life as an archetypal I AM, it will have to make its way through an immense variety of archetypes, many conflicting: "The Good, the True and the Beautiful are clearly universal, and archetypal in so far as they cannot be broken up into prior qualities. But clearly, then, the Bad, False and Ugly must also be archetypal in a certain secondary way."[06,65] Likewise "the Unreal and the Mistaken."

Central in Whitman's thinking is the "archetypal essence or 'reason'" within existences, which makes them intelligible[67,43]. He quoted A.C. Ewing "that there must be necessary connections in the real if we are to apply logic or mathematics to it at all."[67,33] Taking a broad view that summarises much of what is covered in this book, Whiteman wrote:

> The ancient religious philosophies founded on mystical experience—of the Upanishads, Buddhism, Plato, Neo-Platonism, and Christianity up to AD 1300—uniformly taught the knowability of an immense background structure, a subconscious potentiality, a hierarchy of creative powers, a divine Law and Reason (*logos, ṛta*), coming to manifestation on the physical or some other level according to the mind of the beholder, who thus participates in these creative acts. Being the participator, the beholder was also a contributor, both at the outermost level and within, the human spirit being like a focus of powers in the ocean of the universal Spirit. All was thus the outpouring of a Spirit, neither matter nor mind in the limited modern sense of the terms, and deriving from a transcendent well-spring of Unity and Power.
>
> By the "trick" (*māyā*) of creation, the individual knower, patterned on the transcendent I AM, might be caused to *project* and *enclose,* making the

> real appearance of a "world" and a descent of the individual into "matter". Thus came about the deception of dualism of object and subject, matter and thinking. Release was by denying grasping and fixation, passing beyond the sphere of thinking, and thus obtaining the freedom of knowledge of the Spirit at all levels.[67,370]

Consonant with this is Whiteman's rendering of the opening of St John's gospel, cited in Chapter Twenty-five: "Archetypally [*en archē*] there was Reason [*logos*]. Reason was with God, and God was Reason...."[06,66] A quotation from St Paul's letter to the Romans (1:19,20) also refers to archetypal reason, transcribed and expanded by Whiteman as:

> Knowledge of God is apparent [even in those doing ill]. For his invisible [creative] principles, [namely] his everlasting Potentiality (*dynamis*) and [timeless] Divinity (*theiotēs,* "Godhead"), are discoverable [and] understood by the happenings brought about.[06,67]

Commenting on this passage, he wrote, "St Paul, no doubt echoing current teaching in his time, asserts that from created things in the world one can perceive the invisible things of God, namely his 'everlasting potentiality' and Godhead."[06,152] Whiteman's use of the term "Godhead" requires examination in the context of archetypal reason. He noted that in the Upanishads, *Puruṣa* is archetypal God-man, and that "the Unchanging One Supreme" is "clearly a way of describing the Godhead of unchanging universal ideas." He cited Eckhart, "God works, the Godhead does not work, there is no operation in it."[06,152]

Where "archetypal humanity" is concerned, he wrote that "If individual 'I AMs' are admitted, and each is given a meaningful life and a body to carry sense-powers in some kind of 'world', it is surely clear that there must be two kinds of 'body'. There must be one kind that is in perfect or near-perfect *harmony,* as something to aim for; and there must be another kind for individuals who are progressing towards that perfection, and so are at present in some kind of struggle."[06,72] He showed the "struggle" was depicted in many ancient texts, notably in the Old Testament and St Paul, as the corporate personality (the "people") enters revolutionary remodelling in the "day of the Lord" or the "end of days". This transformation, associated with ultimate or near-ultimate contests discussed in Chapter Sixteen, requires careful interpretation. This is the subject of the next chapter.

CHAPTER TWENTY-NINE

DELETION AND TRANSFORMATION

Chapter Twenty-four outlined Whiteman's mystical interpretation of the First Isaiah and the Psalms, which focused not on historical events but on "the loss of security, through an upset of values and personal control, afflicting the prospective mystic prior to disclosure of the divine Source."[06,182] Christianity appropriated and interpreted these and other passages to bolster the idea of Jesus the Messiah, as supposedly prophesied by Isaiah in particular. This interpretation was disputed by Whiteman, who recognised in Isaiah's Book an outstandingly full description of transformation to mystical life by the effective deletion of the old self.

Isaiah's mystical vision of destruction may be clarified by passages translated by Whiteman. In this extract from chapters 2, 3 and 4 of the Book of Isaiah, Whiteman noted that "the three higher levels ('heavens'), to whose influence the individual becomes open, are called by him (in downward order) Zion, Jerusalem ('founded in peace'), and Judah ('praised').[06,110]

> It shall come to pass at the end of "days", that the mountain of the Lord's house [Zion] shall be established at the top of mountains, and exalted above the hills. And all possible human consociations will have streamed to it. Wherefore many "peoplings" will go and say, "Come, let us [go] to the mountain of the Lord ... and He will teach us of his ways ... For from Zion there goes forth divine instruction, and the counsel of the Lord from Jerusalem. But see, [in the "day" of first great upliftment] the Lord of hosts will have taken away from Jerusalem and Judah support, food and drink ... and the "peopling" will be oppressed, man opposing man ... for Jerusalem will have tottered and Judah fallen.

Whiteman interpreted the "hills" as "individuals in ruling identity who are seeking the Right and Good, and the Mount of the Lord is their ideal of Divine Humanity."[06,181] Isaiah then spoke of a "remnant", which Whiteman interpreted as "the remnant of the personality which stays as if grounded in the 'third heaven', while life continues more or less normally in this world (when one is not separated.)"[06,111] His translation continues,

> It shall come to pass in that "day", that the remnant of Israel ["warrior of God"], and they of the house of Jacob ["supplanter" of the worldly self-centred life] who have escaped, shall stay upon the Lord with the certainty of truth. A remnant shall return, even a remnant of "Jacob", to the mighty God. For though thy "peopling", O Israel, be as the sand of the sea, only a remnant shall return. For a final destruction [of the former life-style] is determined, overflowing with sheer goodness and purity.

In Isaiah's great vision, "something like a Last Judgement on his former 'peopling' (or personality) was required, almost all those contributory minds needing to be still fully accommodated to this world, while his core identity remains as if in secret, supervising the whole in the light of guidance from above."[06,111] So, he received the inward instruction:

> Make the heart of this "people" comfortably obdurate, their hearing dulled, and their eyes turned aside, lest seeing with their eyes, hearing with their ears, and understanding with their heart, they return [to full participation in the personality], and so there is [premature] curing [of faults] for them.[06,112]

This surprisingly repressive instruction was interpreted by Whiteman to mean that Isaiah's CMs should remain focused on the physical world, not also raised while spiritually undeveloped. "They must remain with Isaiah at the normal physical level, so that he shall be able to live a normal physical life."[06,112] It is something like pressing the delete key on the old self, but still keeping it in the deleted items file for recall when appropriate.

Similar "divine deletion", release and upliftment appears to have been experienced by Isaiah's contemporary, Hosea. Whiteman translated the beginning of chapter six as:

> Come let us return to the Lord. He has torn us to pieces and will heal us. He has struck us down, and will bind us up. After [those] two stages ("days") he will revive us, and in a third stage he will raise us up, and we shall live in his sight.[06,76]

Whiteman commented that this can be understood in two ways: "(i) As a general principle in spiritual development. First, things must seem to go wrong, so that some degree of discomfort or stress results. Then tolerance and faith are called forth, and finally, an improved skill is granted. More specifically, (ii) the Mystical Way takes us to a phase of continuing 'contest'. We are given the power to face the stress with faith and obedience, over an extended period. Then, in due time, Faith and Obedience being strong, full Release and Transformation, with knowledge of the Divine Source, is granted."[06,179] He also noted that "It is a uniform mystical teaching that disclosure of the Divine Source comes only after the second 'day' of Hosea, that is, with tolerance and abandonment of self-will."[06,112]

In Buddhism, Whiteman noted that "*anatta* [selflessness] must come to rule our lives entirely... Before that completing stage in the Buddhist Path, there is first the stage of 'stream-winning', sometimes described as 'conversion' (which can be understood as finding, and aiming always to follow, the Right and Good). Then, in due time, for those to whom it is appropriate, there can be an intensification of endeavour, after which there normally comes a period of Final Disclosure and Release, *Nibbāna,* in which self *(atta*) gives place to the underlying core identity. That is what happens with a mystical transformation."[06,88]

Whiteman noted that in the Buddhist path, "no reference is made to an Archetypal Humanity (God, Son)."[06,89] In this it differs from other major religions, which can lead to the idea that Buddhism is technically atheistic. Nevertheless, the mystically transformative events of "The day of the Lord" are described as explicitly in Buddhism as in any other tradition. Whiteman pointed out that, as with other religions, "trials, and suffering in general (*dukkha,* in Pali Buddhist teaching), are not wholly bad, to be simply dismissed. They are the means for building up a correspondingly invincible and lasting spiritual power."[06,91] "Building up" necessarily requires a pulling down round "an unwise but initially satisfying path. Suffering comes in when it is time for us to take new stock of our self."[06,163] Then, as Whiteman translates from the Dīgha Nikāya, "the succession of personality-attachments (*jāti*) is finished, the spiritual path has reached its goal, what has to be done is done. There is no continuance of this-worldliness."[06.170]

The Yoga Sutras commence with the statement that yoga is mind-involvement-checking. As Whiteman pointed out, the twelfth *sutra* "makes it clear that the reason why the 'involvements' mentioned need 'checking' is that they may involve us in attachment."[93,163] In the Upanishads, the root cause of attachment is desire (*kāma*); it needs to be deleted, and ultimate transformative release is by "putting-off of deed-attachments (*karma*)."[93,73]

It would be a mistake to assume that the aim of the "mystical way" is then to blot out life in this physical world. It is here "where all spiritual progress is grounded," Whiteman wrote.[06,x] "Mystical evidences or conclusions by mystics can be produced to the effect that spiritual growth towards a final unitive state can occur only at the physical level."[06,33]

This view can only make sense if individual life is seen to extend beyond the physical world; therefore a task of scientific mysticism is to make sense of the idea that there is extension to worldly life. Whiteman's approach to this was both empirical and logical, and a review of his work should end by enquiring how successful he was.

PART V

OVERVIEW

CHAPTER THIRTY

"THE MEANING OF LIFE"

Whiteman's last three volumes carried the title *Old and New Evidence on the Meaning of Life.*[86a,00,06] Looking back over his work, one has to ask what he understood by "the meaning of life", and what we may understand by it. Clues are given in the subtitle of these volumes: *the mystical world-view and inner contest.*

A mystical world-view, Whiteman emphasised, "is no mere speculation. It is testified by a vast body of direct experience. One can trace it particularly in the dialogues of Gotama Buddha, Plato, the New Testament, and Swedenborg."[79,32;00,40] When a mystical world-view is combined with the events of inner contests (Chapter Sixteen), the focus on the meaning of life is strongly on this physical world. Whiteman never downplayed the importance of physical experience and living, and to understand it from the quantum level to everyday happenings. Physical life is the turf for contests involving conflicts and counterfeits (Chapter Sixteen); spiritual development, Whiteman maintained, could proceed only through the trials of inner contests, and the skills that are strengthened by them. Evidently this is turf not readily available outside physical life, which makes physical living crucially important for development, even when viewed as a "vale of tears" (R. Browning). In this he was in accord with world-wide mystical teaching, as noted in the previous chapter.

A private elaboration of this view is contained in a letter written to me on the nineteenth of April, 1992.

> I did have in mind the uniform teaching of the Indian scriptures, as well as Swedenborg, that the "true" disposition of the individual (which distinguishes him from other individuals and is developed in secret, so to speak, beneath fluctuating and temporary influences possessing him) is formed or changed only by *physical* acts—the non-physical worlds provide only a working-out, purification and fulfilment. When imperfect dispositions have been worked out and fulfilled, the springs of life have been "exhausted" and one must return to the physical world in some way for further development. I think this conclusion is based principally on experience in "Separation" (especially of the primary kind) where it is

> quite obvious that when one gets into "the other world" one's power of reflective consciousness and rational choice is limited by what one has attained to in the physical life. One cannot exercise the powers of Recollection there if one has not practised them in the physical life. And when further deficiencies become exposed in Separation, one either passes into a dream-state, or returns to the physical state for further practice there.

It seems one has to be in physical life in order to gain the skills to transcend physical life. The opportunity for spiritual development could be seen to provide the meaning of life in this world, at least in large part. If one does not aspire to spiritual development then life can have little or no inner meaning. A meaning *to* life can be found in many ways—reproduction, the social rewards of family and community, sex, patriotism, sporting, artistic and intellectual pursuits, various ways of self-advancement—but such meaning can be swept away by innumerable hazards and catastrophes. Striving for non-attachment or release from unwise but initially satisfying physically-based living is, in the last analysis, what sparks off development, through inner contests. As T.S. Eliot put it in his *Four Quartets,*

> From wrong to wrong the exasperated spirit
> Proceeds, unless restored by that refining fire
> Where you must move in measure, like a dancer.

The phrase "move in measure" evidently corresponds with the fundamental skill of Obedience in Whiteman's terminology. As Eliot pointed out, non-attachment resembles indifference "as death resembles life"; a meaningful life of non-attachment is played out as a life of positive engagement at the physical level. This is evident in Whiteman's own life, outlined in Appendix B. A main thrust of his work was to encourage engagement in inner contests, stressful though they may be. Physical life provides the opportunity, even though his repeated advice was to be in the world but not of it.

The "meaning of life" may be bound up with seeing the purpose of life (Chapter Nineteen); a purposeless life must seem a meaningless one. It could be said that Whiteman's scientific mysticism contains an antidote to materialism in its form of existential purposelessness. Materialism, according to an "affirmative history and definition" by Richard Vitzthum (1995), recognises "the essential purposelessness and insentience of the material order." (p. 231). It "postulates a single, unified, material substratum or substance as the source and explanation of everything." (p. 18). It is a postulate, a faith; no treatment of materialism has ever passed beyond a postulate into clear demonstration. Expectations for this to happen were reasonably labelled "promissory materialism" by Popper and Eccles (1977, p. 96). As a faith, it may be absorbed unthinkingly, or thought out, or, in the case of Vitzthum, the result of an epiphany.

> Then, one overcast spring day many years ago, I found myself looking up at the clouds overhead and wondering, Could it be that nature is all there is? Could it extend beyond the known universe the way the sky extends beyond those clouds? All I had imbibed not just from religion but from my academic career in literature and from Euro-American culture as a whole had conditioned me to see the world as charged with invisible, transcendent meaning, like a battery charged with electricity.
>
> As an agnostic, I had to grant the possibility that such meaning existed, since plainly I had no proof that it didn't. Moreover, mainline Western philosophy since Plato had argued than an ideal, immaterial realm of meaning not merely existed but in fact governed nature. The possibility that the nature I saw around me that spring day extended outward and inward forever and was thoughtless, random, and impersonal to the core hit me like a revelation. The freedom I felt from the philosophical tradition I'd inherited was instant and all but total. I had become a materialist. (Vitzthum, 1995, p. 16-17).

If nature is "thoughtless", one may wonder what Vitzthum's complex thoughts were doing in it as part of nature, as a materialist must assume they were. There is failure to understand the powers of *māyā* (Chapter Two), but Vitzthum's drive to write "an affirmative history and definition" shows that his epiphany and belief in a "thoughtless, random, impersonal" nature does not necessarily make for personal meaninglessness. His aim (p. 17) was "to write the wonders of materialism." Yet for him the existential situation of people was secondary to purely intellectual satisfaction: "The wide variety of human response to the material reality humans find themselves a part of is less interesting than the reality itself." (p. 232). But if there is "essential purposeless and insentience of the material order" (p. 231), a faith that is centred on this material order rather than on people is not positioned to yield a meaning of life. Ultimately, all is purposeless—including, one has to presume, Vitzthum's book itself.

Vitzthum's account of his epiphany may be compared with Whiteman's "Discovery of The Source and Obedience" while taking a walk "practising Recollection or meditating along these lines."[61,12;00,319]

> Something induced me to look up at the sun, towards the right and about 20 degrees up. And suddenly, but without abruptness, I saw in it the One Only Principle of Loving Wisdom, the Source of All. In a flash, I knew

> that henceforth I had only to obey that Loving Power, and I could not go wrong. All weight of self-decision was instantaneously and completely lifted off, and I became, as it were, a new person, with the loving response and joyful humility I should have had from the beginning, if it had not been covered up by outer darkness and constraint. Though apparently in the physical sun, the Source had an absoluteness of direction which remained with me as a fundamental reality, apart from, prior to, and more real than the physical world.

His "vision of archetypal light" during a separative experience, recorded in Chapter Twenty-six, may also be noted for comparison.[61,35] In this kind of experience, typical of mystics' reports, all is not seen as ultimately meaningless, purposeless. Phrases such as "more real than the physical world" affirm the direct experience of other worlds and other-worldly qualities, which have vast implications to do with the meaning of life and its continuance. To suggest otherwise, Whiteman maintained "offends against the Principle of Sufficient Reason," since it "makes an individual come into existence already furnished with a disposition for good or evil, conferred on him by a 'dice-playing God'."[00,36] As noted in Chapter Thirteen, it follows that "The individual must therefore already have a line of consciousness, with inner memory and corresponding disposition worked out when he/she takes on the new physical life."[00,139] Holding this transcending world-view inevitably brings meaning and purpose into one's life. The idea of its termination at physical death "seems to rest largely on ignorance or rejection of the evidence for 'other worlds' and of processes of separation from the physical body, as well as any logical purpose in regard to the continuing life of individuals."[06,75]

This could be said to encapsulate the mystical world-view, when combined with the necessary inner contests that lead to spiritual development during the brief passage through physical life. Vitzthum's aim early in his book to "write the wonders of materialism" (p. 17) is unable to point to any reason, logical purpose or ultimate source of love and enlightenment in life; only survival in a world that is "thoughtless, random, and impersonal to the core" (p. 17).

Chapter One started with two quotations, from materialism to parapsychology, that took a negative view of mysticism. Husserl also had a problem with it. Whiteman maintained that mysticism belonged to the "highest grade" of phenomenology,[61,40] even though Husserl in his phenomenology regarded the term *mystisch* with disapproval.[61,405] In this respect Husserl was writing in the Kantian tradition, discussed in Chapter Eleven, and seemed unaware that his world-view coincided with the mysticism of classical Indian texts, reviewed in Chapters Twenty-one and

Twenty-two. A revolutionary aspect of Whiteman's work was the attempt to pull the term "mysticism" out of this mire and bring it into the realm of science and phenomenology. One should then expect a "mystical world-view" to be rational and experience-based.

Nevertheless, Whiteman's use of the term "mysticism" could well be a disadvantage in presenting evidence and reason concerning the meaning of life. The common idea that mysticism ignores or bypasses evidence and reason confuses it with religious belief. He kept to the term "mysticism" because it had an established technical meaning in Western literature for "those *higher* experiences in which there is a clear and genuine recognition of the Source, whether or not it is clearly manifested in intelligible sense."[61,119] On intelligibility and religious belief, he wrote,

> It may be that in some descriptions of experience without *intelligible* features, but with strong emotional colouring and religions associations, the name "God" is unsuitably introduced, the subject having formed a wrong belief as to the nature and value of his experiences. Possibly, also, in other experiences, the name "God" is associated in the mind of the subject with theological beliefs which are not warranted by anything in the experience itself, and these beliefs are then misleadingly embodied in the description of the experience. In other words, a theoretical interpretation has been arbitrarily imposed.[61,119]

Whiteman clearly did not consider that mysticism, melding with archetypal reason and intelligibility, had a proper link with formal religion and theology. Standard theological terms were avoided; terms related to direct experience were accepted. This allows the development of a "universal theology", the subject of his last book.[06] Established religions from their side tend to keep their distance from mystics, and Whiteman's views must seem heretical at least to the Abrahamic religions. His views are doubtless also heretical to the confirmed materialist and conventional scientist, which accounts for the word "revolutionary" in the subtitle of the present book: *The Revolutionary Synthesis of Michael Whiteman.* It engages in a revolution that could happen only at the present time, when the formalisms of established religions, classical science and materialist concepts are disintegrating under the emergent or resurgent understanding of consciousness and its place in life and nature. If the revolution seems godless it is because it discards the current lifeless stereotypes of the divine. The meaning of life may now take on a breadth that was not accessible before.

His final statement on the meaning of life may be found in his article on "the mystical way".[86b,654;00,185]

No one who has experienced a real disclosure of the Divine Source and mystical transformation can doubt that this is the goal, purpose, and meaning of all life. Nor can one doubt that if this goal is not attained in the present life, it lies ahead as an effective purpose for the hidden real being of each individual when the outer parts of the personality have partly or wholly fallen away.

CHAPTER THIRTY-ONE

A MYSTICAL LIFE

At the very beginning of his *The Mystical Life*[61] Whiteman showed that entry into the mystical way called for continual discipline, rigorous observation, and determination. His listing of twelve "recollection practices" makes this clear; they call for unbroken dedication to "the assumption of higher attitudes of mind."[61,9] In his own life, recollective practices began about the age of twenty when he entered Cambridge University (Appendix B). His life had already been through some powerful inner events. He recorded that as early as age 5-6, "at night I used to lie, quietly watchful, while the walls of the room receded and dissolved, the same process taking me into a state which I would now describe as dissociated, or incipiently separated in another kind of space. I was fully aware that this other space or state was not the normal state (otherwise how could the walls dissolve?)."

He continued, "I tried to absorb its special characteristics and the way in which I was able to alternate between the two modes of existence, the 'inner' one of which was by no means less valuable or interesting than the 'outer'."[00,41] He was "not engaged in 'thinking'; but I perceived and understood intuitively the character of the situation and phenomena." Thinking "suggests a discursive faculty which must cease its operation in 'that quiet recollection which is the aim of every spiritual person' (St John of the Cross)".[00,43] Self-evident reason is involved, but not discursive reasoning.[61,4] His capacity for "essential insight" and Husserlian phenomenological penetration was evidently being developed at an early age, as was a capacity for Obedience to "a higher power that kept me free of fear, trustfully watchful, and open also to strange indications of non-physical beings watching me."[00,209]

Around the age of 20, a "great discovery" came "when in following some music in the score I suddenly realised that there was a way of voluntarily holding some chosen sound *conceptually* in mind, so that a deeper, precisely characterised and liberating *essence* was revealed in it."[00,315] Following this discovery of "essential insight", "My chief aim became at once to liberate every kind of sensation in that way." The study and practice of music clearly was hugely influential in his thinking.

Development of insight was helped at this age of 20 by a course in Pelmanism. In one exercise "one goes for a walk of about 15 minutes duration, having resolved to 'stop time' on detail after detail, just as they happen to present themselves, and to do this in such a way that one can be sure of being able to reproduce each detail afterwards in the memory with its timeless conceptual-perceptual character."[00,315] This developed a "power of *timeless recall* of particular sensations", which he came to call Recollection. As a Cambridge student, "the dormant faculty of Recollection having been stirred, all that up to now had been wrapped in confusion instantly passed away, and a new space burst forth in vivid presence and utter reality, with perception free and pin-pointed as never before; the darkness itself seemed alive. The thought that was then borne in upon me with inescapable conviction was this: 'I have never been awake before'."[61,57;00,316]

This was before he had become familiar with separative states, but he was aware of having been "vividly located" in non-physical space. After reading Muldoon & Carrington's *The Projection of the Astral Body* (1929), a series of physical-like "duplicate-state separations" were experienced, initially with no bodily characteristics; "I was just a substance located in a non-physical space, separating from another substance somehow identified with the physical body."[00,317] Meanwhile the practice of Active Recollection continued, endeavouring to make the state effortless and continuous. Then came the "momentous discovery, essential for later spiritual development, that such a continuous and effortless kind of Recollection (Continuous Recollection) could be voluntarily induced. [The discovery] expanded into the awareness of a boundless whole whose details were known simultaneously, being open to exploration as on a map without losing primary contemplation of the whole. Time had become like space."[00,318]

This discovery would have fed into his idea of non-physical substructure, especially of time-dimensions. These events around the age of 20, it is worth emphasising, indicate that Whiteman's future mystical development was not based on woolly imagining and vague, undisciplined thinking, as common misconceptions about mysticism might assume.

Having consolidated two of the spiritual skills, Active and Continuous Recollection, the skill of Obedience was consolidated with the discovery of "the Divine Source"[00,319], reported in Chapter Thirty. There remained the question of how the inner revelation and outer physical life could be reconciled, "but a complete and absolute Obedience carried with it the faith—or rather, the conviction of assured knowledge—that whatever happened would be right and good."[00,321] Faith that whatever happened would be right and good was tested by "stresses" which "built up in the solar plexus, spread over the lower part of the body, and reached a kind of limit of bearability before being relaxed. This was followed by the sudden releasing revelation of a divine Fire at the centre of the spiritual body, spreading through the body healingly and awakening other centres."[61,14] This "purging and healing" may have been associated with an activation of the *chakras*, and recalls Pascal's experience of "Fire"(Chapter Twenty-one).

In another "astonishing and indescribable process" there was an "absolute cessation of all, I was merged successively (to outward view) in the Idea of the One, the Idea of Very Self, and then the utterly all-sufficing Idea of the Good, nothing whatever existing then (if such inadequate words can be permitted) except those unchanging and timeless Ideas in God."[61,14]

These reports are not likely to convey much to many readers, although parallels exist in various sources, as in the fourth Buddhist *jhāna,* and *kaivalya* or oneness in the Yoga Sutras (Chapter Twenty-one). But they give some idea of the experiencing of universality, which would flow into his ontology. He became concerned with "the difficulty in producing a verbal description that might awaken in readers' minds some true conception of the experience which it is meant to describe".[86a,135] All he could "reasonably hope for is a gradual awakening of the reader to conceptions which he does not already have clearly within him."[86a,135] For his own part, experiences like these evidently contributed to his handling of terms such as "archetypal" and "universal".

Around the age of 25 inner events were accompanied by psychological disturbances such as dissociated states, and seemingly bizarre behaviour like buying apples and throwing them across the street, which he took to be a test of obedience. These "symbolic performances ... must have been very disturbing for my family, but which I felt 'the Word of the Lord' forbade me to discuss with them."[00,333] He recognised that these ordeals "would be more comfortably resolved if I could follow the practice of Indian seers and some Christian mystics (e.g. St Paul) and retire to a forest or desert."[00,332] Yet progress along the "mystical way" was not seen by him to need "guidance from without as to what behavioural patterns may be appropriate and when or how to practise particular techniques of self-discipline and release. The 'inner voice' will give all the guidance that one needs; as classical mystical teachings declare."[86b,633;00,161] He added, "It is all too easy for many people to obey the injunctions of an 'outside' teacher, thus relieving themselves of the more stressful (and less showy) endeavours needed to rectify their adverse motivations through perseverance in *inner* contests, with trust in *inner* teaching."[86b,633;00,162]

Even so, without physical help or guidance the outcome was a brief period in a mental institution "in a deeply dissociated or separative state, dealing with spirits or receiving instruction."[00,344] Instruction concerned "unfamiliar spiritual phenomena and skills", but at other times there was "the obscure but deeply troublesome presence and activities of adverse entities."[00.336]. This was evidently a time of ultimate contest, as described in Chapter Sixteen; it may be regarded as a kind of shamanic initiation, but despite uncomprehending psychiatric handling, it was "an amazingly unbroken time of enlightenment and empowerment especially in regard to inner potentially disruptive forces." (Letter to me of 31/12/2001.)

On recovering, "I knew that the former stabilised contest-dissociation was over, and that a healthful higher dissociation, able to interfuse the physical state with perfect harmony, was in full control."[00,338] The remainder of his long life showed completely "normal" behaviour and competence, although aware of being "in the world but not of it".

His diaries contain accounts of over 7000 psychic, separative and mystical experiences. He did not cultivate psychic ability, but spontaneous precognitive experiences had an important influence on his recognition of different time dimensions (Chapter Nine). No comprehension of his world-view is possible without recognising the multi-level and multi-world insights that flowed from his own direct experience in a mystical life, together with confirming sources ranging from ancient classics to modern physics and Husserlian philosophy.

For Whiteman, a mystical life did not mean disengagement from living in the physical world. His output shown in the References below is that of a prolific writer, even though the list is selective and does not include a large output in the field of music. Nevertheless, his published work is in a sense the cream atop a huge quantity of rejected submissions to book publishers and journal editors. His correspondence shows frequent frustration and dismay at the obstacles and incomprehension that rose up continually against his work. He put it down to editors and reviewers being unable to free themselves from the pervading one-level physical naturalism; on October 10, 1989 he wrote to me that "I sometimes feel 'lonely' in the present world-outlook, with publishers and hordes of scientists not seeming to understand what I try to say, whether about Recollection, or about modern Physics, let alone mysticism."

As a mystic he saw teaching and instruction to be important. In 1970 he gave a course of nine lectures on mysticism at the University of Cape

Town Summer School, and conducted private study groups in mysticism, Sanskrit and Biblical Hebrew. He saw engagement with these activities to be a necessary part of the mystical life, since,

> when the mystical power begins to fall off, there comes the knowledge that only by *going out* can one satisfy Obedience and eventually become stronger to remain close to the Source. Hence the "western" power, which means a setting of the Divine Sun, must be permitted stronger expression, for the purpose of outward particularization and grounding."[61,233]

Nevertheless, he noted a "danger, on thus going out, of a complete lapse into 'the world', with loss of even the secret knowledge of the Source."[61,234] No doubt this applies to most of us. "The world is too much with us," Wordsworth lamented.

What of a call to a mystical life for other people? Whiteman's view was:

> Whether, then, the manifestation of inner reality comes to us before or after death seems to me a matter of relatively little importance to the individual, unless a definite need exists. For what is important is the potentiality of the individual for good; and the growth of that potentiality, while dependent in one way on a certain manifestation of inner reality, is also, in another way, independent of such manifestation. And the outcome will not be finalized for anyone till his work on this lower plane is done.[61,21]

CHAPTER THIRTY-TWO

SUBSTRUCTURE: BED FOR A HARMONIOUS TRIO

At the end of his last book, Whiteman posed the question raised at the beginning of this book: "What has Mysticism, which is purely other-worldly, to do with Physics, which deals only with this world?"[06,313]

Yet the word "purely", he noted, begs the question. If one-level naturalism is discarded, a mergence between mysticism and physics becomes evident through their connection in substructure. Mystical teachings, Whiteman pointed out, "extend into kinds of experience and creative cycles which lie causally hidden, as substructure, in all existence and human awareness." Likewise the advances of modern physics also extend into substructure, shown by "the metaphysical insights of tensor analysis and other elaborate mathematical structuring, as in electromagnetic fields and their interaction."[06,313] The same role played by substructure in connecting mysticism and physics also makes the phenomena studied in psychical research accountable. As discussed mainly in Chapters Two, Three, Four and Ten, a substructural nexus can be said to provide a bed in which science, mysticism and psychical research all harmoniously lie.

Even so, Whiteman noted, "it may be complained that the physical and the mystical are so different that, in order to avoid confusion, we are obliged to study them as in different departments, so to speak."[06,313] His answer was "to show how, beginning with mystical experience and metaphysical principles, not only is the tensor basis of the Theory of Relativity derived (this has already been done, by Einstein and others), but also the highly complex structuring of Quantum Field Theory becomes a matter of metaphysical necessity, at least in its broad foundations and basic formulae."[06,314]

"The first major truth" regarding this commonality, he wrote, "is that a 16-fold creative cycle, with an infinite variety of developed forms and applications, stands at the root of all existence."[06,314] The conjunction of objective and subjective cycles was discussed in Chapter Ten. So he wrote, "A fact that must not be overlooked, if the actual world of Physics is in question, is that two cooperative kinds of 16-fold cycle are required.

On the one hand there must be cycles which apply in a universal way to all individuals able to discern what is going on. These objective cycles are concerned with the primary qualities of Galileo (space, time, mass), and must be logically precise and certain. But nothing definite is observed unless there is combination with a subjective cycle determining the observer's response."[06,315]

The subjective cycle needs to be seen in its full range of physical and non-physical experience. We should avoid "arbitrarily making a rift between physically measured space (the results being then always approximate and doubtful) and the various kinds of mental or spiritual space (dream, imagery, psychical space, 'out-of-body' experiences, mystical transport and transformation, and transitions between all these). We shall be very deficient in understanding if, while being expert in physical space-measures, we are totally at sea in explaining other worlds of space and time (dream, psychical, mystical)."[06,313]

Space and time have figured prominently in this book. Whiteman noted that the most obvious factors in existence are "Objectivity and Change", which involve space and time respectively. For, "it is difficult to see how there could be any kinds of objectivity without some kind of space to 'hold' the various objects of our attention. It is also difficult to see how there could be change without some kind of time, to 'hold' the various stages of existence of what is otherwise to be regarded as 'the same' in some logically prior way."[06,316]

In thinking about space and time, one needs to recognise the joint operation of objective and subjective cycles. Time might seem to go fast or slow, space might seem to be short or long, yet "meanwhile the exact cause is present as a *substructure,* seemingly in whatever is observed."[06,317] Substructure underlying individual life was seen to be patterned on what Whiteman termed "Archetypal Humanity", "covering all discernible causes in human existence and behaviour, free from counterfeits. This, along with counterfeits corresponding to the stage of development reached, will provide the pattern for each human individual, leaving scope for differences in *personality* to be manifested in various ways and at various times while *identity* remains the same."[06,319]

In the case of substructure underlying observed objects, some discussion was given in Chapter Two. It was noted that when several people are observing the "same" object, their experiences will be different—different perspectives, different lighting and so on. Yet the observations must be logically compatible; "the measures of the same object for another observer in the same world must be *transformed* according to some rule."[06,319] The implications of transforming rules are larger than one might suppose: Whiteman asserted that "From this simple requirement (worked out in Tensor Calculus), along with properties of the 16-fold creative cycle, result the major field laws of Physics, gravitation, the electromagnetic equations of Maxwell and Dirac, and quantised interaction between such fields."[06,320]

The mathematics intended to show how this comes about were given most fully in the supplement at the end of his last book (2006, pp. 321*f*), reproduced here in the as Appendix A. Mathematical treatment was also given in 1975a, pp. 193*f*, 1977a, pp. 741*f*, and in 1986a Chapter 14, where the three time-like variables (T, τ and t, Chapter Nine) were included. "Obviously," Whiteman wrote, "this is only the beginning of what could be a vast field of research."[94,84] One conclusion involving substructure was that "all creation in a world will be normally wavelike at its creative roots. But it must be understood that all this is in *potentiality,* for rays in all directions at every point. Nothing in particular would be seen without *subjective* cycles which manifest, for some individual, all the potentialities that are relevant to his/her mental state and capacities at the time and in the world in question."[06,324] Such, in Whiteman's view, is the connection between physical laws and the life of an individual, "through whom the knowledge becomes definite."[06,331]

The final conclusion of his last book, then, is again to recognise "Galileo's distinction between primary qualities (space, time, mass) and the secondary qualities, which are actually observed according to the individual's position, time and mental powers."[06,331] To arrive at a synthesis of what we experience in the physical world, we need to recognise "a *subjective cycle,* such as mystics, many theologians, and some psychologists have admitted. This cycle in its most complete form is sixteenfold, and parallels exactly what may be called a 'world cycle' (as it appears in quantum field theory) in every way, while manifesting itself in every kind of human experience."[06,331] In this book the idea was outlined in Chapter Ten. The combined cycles can be seen as "a form of potentiality which is discernible, and indeed can be *lived* by the individual in an enormous variety of 'altered states of consciousness', revealing, to the very life, what was previously in the physical world as only a surface acquaintanceship and composite adaptation of feeling and motivational states, themselves physically unlocatable."[06,332]

Yet such substructure, he wrote, "can be 'lived' only if the individual (in a secret *core identity*, or other inner identity, as in dream) passes into another world. And this requires also transitional states leading to worlds which are (representatively but really) governed by corresponding variants of the *objective* sixteenfold cycle studied here." It requires a phenomenological approach initiated by Husserl, discussed in Chapters One, Eight, Nine, Twenty-one and Thirty. In this approach "the physicalistic hypotheses are bracketed off, and through a suspension of attachments, and a stoppage (epoché), we come to a 'unique form of consciousness' which provides the 'unshakable because self-evidencing conviction of truth'."[06,332] It is to "take stock of the whole picture, from the timeless font of all Ideas and actions of the spirit, to the responses of the located individual through whom the knowledge becomes definite."[06,331]

It is knowledge that reaches beyond the boundaries and marker stones declared by Kant (Chapter Eleven) and like thinkers. It has to do with what Jung (1963, p. 300) recognised as the "decisive question for man," which is, "Is he related to something infinite or not? That is the telling question of his life." While writers like Vitzthum (1995, Chapter Thirty) find satisfaction in a negative answer, there are many, Jung included, who feel liberated by the knowledge of space and action beyond the physical world. More than that, it allows deeper insight into human behaviour and the human condition. I hope that what has been covered in this book shows this to be a justified and sensible position to take.

APPENDIX A

THE MYSTICAL DERIVATION OF PHYSICAL LAWS

This Appendix is part of Supplement A of Whiteman's *Universal Theology and Life in Other Worlds* (2006), which is Volume 3 of *Old and New Evidence on the Meaning of Life: The Mystical World-View and Inner Contest,* pages 320-331. The term "mystical" refers to the identification of the "creative roots" in a world as potentiality, or "substructural space".

Communal Space-Time Measurement; Tensors and the Origin of Physical Laws

Philosophically, the most important contributions to an understanding of modern Physics are, arguably, the Tensor Calculus of G. Ricci and T. Levi-Civita (1901), and the little-known work of H.P. Robertson (as in the *Jubilee of Relativity Theory*, Basel, 1956 pp. 128f), which explores the very basics of measurement in the physical world, on the ground of Group Theory. Some attention must be given here to leading ideas and terminology in all this work.

Looking to the origin of all possibility of communal measurement, we must first admit the idea of an "observer gas", each point being occupied by a mythical observer with a "standard clock" recording isometric time intervals (deviations from this *presuppose* what is isometric). Light rays with a velocity beyond the possibility of change in direction or magnitude must also be postulated; for any change would involve some prior cause already measured. And mirrors must reflect light to other observers with the same speed, because again a change of speed would require a prior cause.

From the various times to and from a distant event, Robertson constructs a *cosmic* time $d\tau$ and a *cosmic* distance $d\sigma$, along with a *metric* for the cosmic *interval* ds between two near events. This extends

Pythagoras's theorem for three spatial intervals dx, dy, dz, squared and added, along with the negative of the squared cosmic time interval, the result being - ds^2. The main conclusion is that time can be grouped with space in all operations if the symbol i (with i^2 = -1) is attached to each time measure, along with squares of the spatial intervals.

Another conclusion is that although distant galaxies seem to be receding (if we apply classical formulae to their spectra), a factor varying with time could appear in the metric. So it could be that the scale of measures adopted for events and the observer is free to vary slowly in time. But the universe would remain finite and unbounded, and no so-called "expansion" would occur.

The laws of nature must be valid for every observer in our "observer-gas". This imposes enormous restrictions. It is hardly ever realized, for instance, that if a man in a train with velocity V is moving forward with velocity v, his velocity relative to someone at rest outside the train cannot be exactly $V + v$. For the modes of measurement in the two cases are quite different. Tensor calculus explores the conditions under which groups of measures ("components" in various directions) constituting a field are subject to *transformation* when another observer measures them. For two people looking at an object never see quite the same thing. Yet we regard them as seeing the same thing (in another sense of the term "same").

If now a physical phenomenon is at (x, y, z, t) for one observer, and at $(\bar{x}, \bar{y}, \bar{z}, \bar{t})$ for another, we must have relations such as $\bar{x} = f(x, y, z, t)$. But only the changes in some small "focal" region call for attention now, by writing: $d\bar{x} = dx\frac{\partial f}{\partial x} + dy\frac{\partial f}{\partial y} + \ldots + \ldots$ There will be 16 partial derivatives, which we can arrange in square matrix form, with four rows and columns. And if we take (dx, dy, dz, dt) as a "row", multiply the terms in this "row" with each of the terms in each of the four columns and add them, we obtain the correct result. The same matrix will now apply to a *field* $F = (F_x, F_y, F_z, F_t)$ for transformation to $\overline{F}$; for the field at any point and instant is represented in components in exactly the same way as a small displacement at that point. For a 2-rank tensor, conveniently written F_{ab}, the transformation formula will be as follows (adopting the Einstein convention, that a repeated letter in suffixes must be summed 1-4: and each T is a 16-term matrix, with four rows and four columns):

$$\breve{F}_{i\kappa} = F_{ab}T_{ia}T_{\kappa b}\text{, where } T_{\kappa b}\text{, for example, means } \frac{\partial \bar{x}^{\kappa}}{\partial x^{b}}.$$

The Six-Dimensional Tensor-Calculus and Sedenion Algebra

Let us now see what happens when we introduce two new variables (τ and T, numbered 5 and 6) which are not measurable, and so cannot be included in the transformation. The field tensor is 6-dimensional, with 36 elements. But a general "square" lay-out of measures can always be seen as the sum of a symmetric lay-out and an antisymmetric one. The first of these kinds would be illogical, since some elements would appear twice, and others (in the diagonal) once. In the case of an antisymmetric tensor, however (with zeros along the diagonal, because terms there cannot be duplicated $\pm$), there would be 15 different elements. And, as shown below, a scalar term must be admitted because one kind of product of two such tensors (necessary for development to be possible) is a scalar. Here we have a pattern of 15 + 1 constructive stages, exactly as in the "subjective" 16-fold cycle.

When we come to the transformations required if the lay-out is to represent a tensor, the fact that τ and T cannot be included in the transformation has remarkable results. Firstly, elements in the 5th or 6th rows and columns will have only one transformation matrix, and so will be vectors. But the element F_{56} will have no transformation matrix, and so will be invariant. Furthermore, if we ignore transformations involving t (i.e. consider only the state of the field at a given time), the first three elements in the 4th column constitute a vector with three components. For the transformation matrices reduce to 3^2 ones, all the other elements in them being zero. We shall see also, from formulae that result, that the elements F_{23}, F_{31}, F_{12} also form a three-component vector when t does not change. We thus obtain an extended electromagnetic tensor as follows (the directional operator i has been inserted in order to conform with Special Relativity in treating t along with x, y and z):

$$F_{ab} = \begin{bmatrix} O & iH_3 & -iH_2 & E_1 & J_1 & G_1 \\ & O & iH_1 & E_2 & J_2 & G_2 \\ & & O & E_3 & J_3 & G_3 \\ [negative & & & O & iJ_4 & iG_4 \\ reflections & & & & O & iM \\ here] & & & & & O \end{bmatrix}$$

It will prove enormously helpful in operation with these tensors if we show the position of a component in the lay-out by attaching the symbol σ_{ab}, where 'a' is the row-number and 'b' the column number, to the respective component, the scalar being σ_{00}. Then the field $F = \sigma_{ab}F_{ab}$ (summation for a and b implied). To describe how the field changes according to some natural law (as in electromagnetic theory) we must introduce a *gradient operator*. In two dimensions, at any point (x, y) of a curve, which we can think of as a hill, the gradient is dy/dx at the point in question. In general, we must beforehand make sure that all operations maintain the tensor character of what they operate on. So we must attach σ_{ab} symbols denoting the first four places in the fifth column (which we have seen to determine a 6-vector, if no other places are occupied). Thus the tensor gradient operator is

$$\nabla = \sigma_{15}\frac{\partial}{\partial x} + \sigma_{25}\frac{\partial}{\partial y} + \sigma_{35}\frac{\partial}{\partial z} + \sigma_{45}\frac{\partial}{i\partial t}$$

The result of operating on a tensor field will then be a tensor if we obey the rules for obtaining an anti-symmetric product from to given ones. There are three kinds of such products: the scalar product, the cross product, and the vector product; and the rules for obtaining these, in terms of the σ symbols attached to their components are:

$$\sigma_{ab}\sigma_{ab} = 1, \ \sigma_{ab}\sigma_{ac} = i\sigma_{bc}, \ \sigma_{ab}\sigma_{cd} = \sigma_{ef}$$

(if $abcdef$ is an even permutation of 123456)

(NB: No summation here)

The Field Equation

The simplest field equation will result then from putting the product of the gradient operator (as above) and the field tensor equal to zero, while J and G vectors and the invariant M are also made zero. The result is then Maxwell's famous equations for the electromagnetic field in vacuo.

Thus the terms in σ_{54} are only $\frac{\partial}{\partial x}F_{14}+\frac{\partial}{\partial y}F_{24}+\frac{\partial}{\partial z}F_{34}$, making up what is called the *divergence* of E (div E), a measure of the "expansiveness" of the field. This must be zero in the simplest case. Similarly div H = 0 in the simplest case. These are two of Maxwell's equations for the vacuum. Two other equations determine the rotational character of the field, with measures curl E and curl H. The first of these turns out to be the rate of change of the magnetic field. This shows how, in a coil of wire, an electric voltage is induced when a magnetic field around it changes (as in a dynamo). The reverse process occurs in an electric motor. The second equation is similar.

It is possible to construct a 4-vector **A** from these equations, and then derive for it a "wave equation". If **A** is constant, this describes a wave traveling forward with steady velocity in that direction. So all creation in a "world" will be normally wavelike at its creative roots. But it must be understood that all this is in *potentiality*, for rays in all directions at every point. Nothing in particular would be seen without *subjective* cycles which manifest, for some individual, all the potentialities that are relevant to his/her mental state and capacities at the time and in the "world" in question.

Obvious means for developing the theory now follow. Thus we can add to the gradient operator a vector field (e.g. the field **A** whose existence we have deduced) and a constant. But we must ensure that the *dimensions* are right. For instance, for t we must write ct where c is a velocity, so that the whole has the same single space dimension as x. The result, with constants agreeing with experiment, is Dirac's equation (here for any field except gravitation, which is explained another way):

$$\left[\hbar\left(\sigma_{15}i\frac{\partial}{\partial x}+\ldots+\ldots\right)+e/c\left(\sigma_{15}\mathrm{A}_x+\ldots+\ldots\right)-mc\right]\mathbf{F}=0 \quad \ldots..(1)$$

where $\hbar$ is Planck's constant, e and m are respectively the charge and mass density in the field, while c is the velocity of light in vacuo. Particles are no more located in the field than the sound of a trumpet is located in an airwave to the ear. A particle-like effect comes only when two fields interact.

Wave Fields and Classical Dynamics

Accepting that the logical basis of all observable changes in this world is wave-like, the next step is to show how wave-like structures can correspond with concepts of mechanics and their measurement in this world. The mathematics is abstruse; but the most important conclusions can be outlined in strict logic and understood by anyone familiar with the notation and chief theorems of first-year calculus. For others, there are general conclusions of enormous importance for a philosophy of life.

We must first be clear about the meaning of an *eigenvalue equation*, of which a very simple example would be $H\psi(x,y,z,t) = E\psi(x,y,z,t)$, where H is the energy operator and E the value observed for a certain $\psi(x,y,z,t)$ (state of the system at the time). The "state" may be made up of numerous (or an infinity of) sub-states, as in the following formula for a Fourier Transformation to a wave-function in two variables x,t from an infinity of waves in what may be called "substructural space". For the separate waves cannot be directly measured physically, though their contribution is essential. In this formula the variable x indicates an abbreviation for what should be three spatial variables (x,y,z):

$$\psi(x,t) = \int \phi(p,E) e^{i(px-Et)/\hbar} dpdE$$

This kind of formulation may be likened to the analysis of a musical tone into its various harmonics, none of which can be clearly distinguished in a normal physical state. The harmonics can be heard separately only by becoming partially "dissociated", blotting out the totality of physical sound, and isolating a particular harmonic (in a "sub-structural" space-time, in which one can stop and analyse what one hears).

In the above equation $\phi(p,E)$ is the amplitude of a wave whose velocity and frequency are given by multiples of p and E (More strictly, if the formula is to be reversible, a numerical multiplier must be placed in front). We see at once that by differentiating under the integral sign $-i\frac{\partial}{\partial x} = p$, and $i\frac{\partial}{\partial t} = E$, on the strict understanding that the *differentiation* applies to ψ, while the measure *multiplies* that from which it is transformed (ϕ).

The measure p so found is usually described as the "momentum" of the field at that point and time (strictly a density), while E is its "energy" (or energy density). Commonly the values found are attached to a supposed "particle". These characterizations now need to be explained. But the mathematics is very abstruse and mostly only general effects concerning our world-view can be indicated here.

Hamilton's Equations

In advanced classical dynamics it is shown that if a "system" is considered as having "states", each determined by a finite number of "coordinates" q (as would be the case with a number of rigid bodies in a "field"), it follows logically that the motion, in accord with Newton's laws, will obey Hamilton's equations:

$$dq / dt = \partial H / \partial p \text{ and } dp / dt = -\partial H / \partial q$$

where $H(q, p)$ is a certain function in terms of coordinates q and p, consisting of the total kinetic energy T of the system along with a term depending on the external force-field. The remarkable fact, now, often overlooked, is that these very equations can be deduced in Quantum Theory, for "expectation values" $<q>$ and $<p>$ calculated from a "probability density" in terms of ψ, as above. In the classical analogue, p stands for a Hamiltonian *momentum*, which, in the simplest case is mass times velocity for a particle. We need to note also that "angular momentum" is the name then given to the "moment" of momentum, which is the turning effect of the momentum of a point in the field about the origin of coordinates. But the analogy with particle motion breaks down at nuclear and even molecular levels since, by the uncertainty principle, precise and sure measurement cannot be made there, and the derivation of Hamilton's equations in expectation values breaks down. But we may expect some correspondence between effects in quantum theory and classical picturings, as substructure – not in *physical* space-time. For because the analogy breaks down at atomic levels, we must not picture events there as if in normal physical space-time. Today, they are commonly spoken of as subject to "non-locality".

The quantisation of fields

We now have to see how quantisation arises. This is a very complex area of research, unfortunately requiring a grasp of some heavy mathematics; so only an introductory outline can be presented here. Anyone not familiar with the basic symbolism is recommended to devote some time to understanding at least the foundations of the calculus (differentiation and integration) and the use of the symbol i $(i^2 = -1)$ to permit space variables x, y, z and time t to be treated on the same footing, as Special Relativity demands.

We begin here with a remarkable piece of work set out in Dirac's book (1930, 1982, pp. 80-91). There it is shown that if there are algebraic symbols q and p such that $qp - pq = i$, then the only possible values of $\left(p^2 + q^2 - 1\right)/2$ are 0,1,2,3, … This expression in p and q is the very same as one that appears in Quantum Field Theory, and is called N. That theory is the result of applying, to any dynamical system, an equivalent of the (Fourier) analysis of musical sound into distinct harmonics – a continuous effect being thus shown to be equivalent to a series of discrete terms added together.

As in the production of a musical sound, a wave-like influence is thought of as carried through a "field" $\psi(x, y, z, t)$, or for short $\psi(q, t)$, where q stands for any number of positional coordinates q_1, q_2, etc. And each "harmonic" is taken to be the product of a time-dependent factor; $a(t)$ and a space-dependent one $u(q)$. Rates of change at any point are given by $p = -i\dfrac{d}{dq}$ operating on ψ. (If we try to correlate with classical dynamics and q is a distance, the scale factor $\hbar$ must multiply the right-hand side.)

We now treat the field as made up of "harmonics", as with musical tone, the whole ψ being made up of contributory $\psi_k (k = 1,2,3,\ldots)$, each corresponding with a "harmonic" chosen. And for each such "harmonic" we distinguish a part depending on t, namely $a_k(t)$, as multiplier of a part depending on the position, $u_k(q)$, where q could stand for x, y, z or any other space variables.

The next step is to define

$$a_k = (q + ip)\,;\ a_k^{\ *} = (q - ip)\,;\ \text{and}\ a^* a = N\,,$$

for each "harmonic" in the field. It turns out that for each subsidiary N in the whole N, we have: $N_k = (p_k^2 + q_k^2 - 1)/2$.

Furthermore, if we operate on the whole field with some particular a_k the value of N for that "harmonic" is reduced by 1; and if we operate with a_k^*, the value goes up by 1. In a slight modification of this plan, N can assume only the values 0 or 1. This conforms with the experimental conclusion that there are two kinds of quantum statistics – light, for example (with no mass-structure), obeys the reduced possibilities of Einstein-Bose statistics.

Angular momentum and particle spin

Here we need to explain effects for which the customary explanation is in terms of what is called "electron spin", or "isotopic spin" of a nucleus, sometimes said to be in "systems of identical particles" observationally indistinguishable (Messiah, 1962, Ch. 13). The mathematical character of such "spin" is inferred from experimental results. Here we point to a certain oversight concerning what is called "angular momentum". Remedying this explains the anomalous phenomena without introducing any kind of "spin".

First we must look at a formula for the effect of rotation of axes in sedenion theory, i.e. with anti-symmetrical 2-rank tensors.

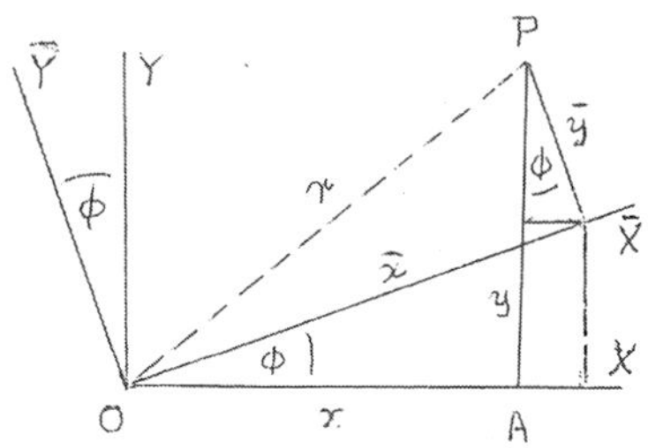

It is easy to see from the accompanying diagram that upon rotation of axes from X to Y clockwise about the Z axis (pointing forward) the following equations (simplified for very small angles) hold:

$$x = \bar{x}\cos\phi - \bar{y}\sin\phi = \bar{x} - \bar{y}\phi$$
$$y = \bar{x}\sin\phi + \bar{y}\cos\phi = \bar{x}\phi + \bar{y}$$

These results are arrived at if we place multipliers before and after the coordinates x and y with their directions σ_{15} and σ_{25} as attached, as follows:

$$\left(1+ \tfrac{1}{2}i\phi\sigma_{12}\right)\left(\sigma_{15}x+\sigma_{25}y\right)\left(1-\tfrac{1}{2}i\phi\sigma_{12}\right) =$$
$$\sigma_{15}(x+y\phi)+\sigma_{25}(y-x\phi)$$
$$= \sigma_{15}\bar{x}+\sigma_{25}\bar{y} \qquad .....(2)$$

Instead of the position vector (x, y) in this equation we can now take the components of a field (F_x, F_y), and the same formula will hold. Looking back then to Dirac's equation, formula (1) above, it is clear that **F** must have 16 components, and there will be 16 equations to solve. But following Dirac, we can introduce a representation of each σ_{ab} as a 4^2 matrix, operating on an **F** with *four* components (and called a spinor). A set of such matrices, with the properties of our σ_{ab} is not difficult to produce. One such set is given by Eddington in his *Fundamental Theory* (1949, p.142; Eddington's $E_{ab} = i\sigma_{ab}$).

Thus: $$\sigma_{15} = \begin{matrix} 0 & 0 & 1 & 0 \\ 0 & 0 & 0 & -1 \\ 1 & 0 & 0 & 0 \\ 0 & -1 & 0 & 0 \end{matrix} \qquad \sigma_{25} = \begin{matrix} 0 & 0 & 0 & -1 \\ 0 & 0 & -1 & 0 \\ 0 & -1 & 0 & 0 \\ -1 & 0 & 0 & 0 \end{matrix}$$

Equation (2) will now take the form SFS^{-1}, where **F** (now represented as a 4^2 matrix) can be taken as the product of two 4-vectors, giving $S\psi_a\chi_b S^{-1} = 0$. This is a limitation, but at least it presents the possibility of simpler solutions. For it amounts to two equations: $S\psi_a = 0$ and $\chi_b S^{-1} = 0$, the second of which is the conjugate imaginary (i changed to $-i$) of the first, and so can be discarded. Thus we now have four equations, describing a four-component spinor field ψ_a. It should be noted, however, that there is no reason to take the components of a spinor as three of space and one of time. Like the symbol i, a spinor is no more than a useful mathematical artefact. There is no ground for treating it as a physical existence.

The so-called *angular momentum* operator (taken to refer to rotation about the Z-axis, when spherical polar coordinates r, θ, ϕ are used) is $M_z = -i\hbar \frac{\partial}{\partial \phi}$. Its eigenvalue $\overline{M}_z$ for a field **F**, given in terms of those coordinates (but having x, y, z components) is $\overline{M}_z \mathbf{F} = -i\hbar \frac{\partial}{\partial \phi} \mathbf{F}$. It will appear, however, that if that formula is accepted as it stands, the inevitable effect of the change in the components of **F** due to the rotation (of axes or field) that occurs when the angle of rotation ϕ is changed has been ignored. So a certain addition must be made, at least in fields described in terms of 2-rank tensors.

Looking back at formula (2) and reading $(F_x, F_y) = \mathbf{F}$ in place of (x, y), we see that the term required (additional to $\overline{M}_z$ as above) is $-i\hbar(\frac{1}{2} i d\phi.\sigma_{12})$, that is, $\frac{1}{2}\sigma_{12} d\phi.\hbar$, operating on **F**, *in addition* to what is normally accepted; $\left(-i\hbar \frac{\partial}{\partial \phi}\right) d\phi$. The addition to the angular momentum operator is $\frac{1}{2}\sigma_{12}\hbar$. And the eigenvalue s of σ_{12} is found from the equation $\sigma_{12}\psi_a = s\psi_a$, where ψ_a relates to rotation ϕ about the Z-axis.

Multiply each side of this equation by σ_{12}, recalling that $\sigma_{12}\sigma_{12} = 1$. It follows then that $\psi_a = s^2\psi_a$. So $s = \pm 1$, and the term to be added is $\pm \frac{1}{2}\hbar$, in a direction *arbitrarily chosen* for our Z-axis (see Paul Davies, 1985, pp. 32f).

One can hardly imagine a more telling refutation of the view commonly held that in all fields there are localized "particles" with measures objectively given in nature, but at the same time such that they vary according to how each individual observer has chosen to place his axes.

It is also stated (op.cit., p.82) that "particles are known" with spin 1, 1 ½ or 2. This, if the experimental conclusions are justifiably drawn, could easily be explained as an effect of *superposition*, just as the multitudinous sounds given out by an orchestra in a single sound wave to the listener can yet be analysed and vividly heard singly in a "higher dissociated" state. Accordingly, we have only to take different values for e and m in Dirac's equation (with possibly also different ranges of vibration-frequency) to obtain what could be quite a different field, both in its creative origins, and in the mind of the observer.

APPENDIX B

A BRIEF BIOGRAPHY

Figure B-1Michael Whiteman aged 84. Sibyl Morris

Joseph Hilary Michael Whiteman was born in London on the 2nd of November 1906. His father, under the theatrical name Sydney Carroll, was an actor, impresario, publisher, editor, theatre and film critic. He founded the Regent's Park Open Air Theatre together with Robert Atkins and "discovered" Vivien Leigh among others. Michael, the youngest son of Sydney's first marriage, was educated at Highgate School, London. He won a scholarship to Cambridge University (Gonville & Caius College), where he studied between 1926 and 1929 to obtain a first class in the mathematics tripos. It was at Cambridge that his mystical development began (Chapter Thirty-one).

After a spell in his father's publishing house he became scholastic head at Staffords School in London's Harrow Weald in 1933. There he met his musician wife Dorothy Eavestaff, known as Sona, a pupil of the famous piano teacher Tobias Matthay. The couple emigrated to South Africa in 1937, settling in Cape Town, where he had been appointed to the Diocesan College (Bishops). In 1939 he was appointed a junior lecturer in the Department of Pure Mathematics at the University of Cape Town. His PhD thesis of 1944 was entitled *The Foundations of Mathematics.* He was active in music, started *The South African Music Review,* was involved in composition, conducting, orchestral performance, and in 1941 took on editing *The South African Music Teacher,* a position he held for 55 years. His editorials are outstanding and deserve separate publication.

He was three times president of the S. A. Society of Music Teachers, where he was held "in high regard and affection". Between 1938 and 1947 he was awarded Trinity College diplomas in composition, a B.Mus. in the University of South Africa, and M.Mus. at the University of Cape Town.

In 1944 he moved with his wife and daughter, Sibyl, to Rhodes University, Grahamstown in the Eastern Cape, where he was appointed a lecturer in music. His wife was principal piano teacher at the university. But, on his appointment as lecturer in the Department of Applied Mathematics at the University of Cape Town, they returned to the "Mother City" in 1946. He lectured principally on relativity, quantum theory and related subjects. In 1962 he was appointed Associate Professor of Applied Mathematics at the university. On his retirement in 1972 he was given the title of Emeritus Associate Professor of Applied Mathematics.

All this time he continued being active in music, composing and preparing students for theory of music examinations. He founded and conducted amateur and youth orchestras, and arranged and composed music for them. He conducted his own piano concerto in the Cape Town City Hall with his wife as soloist. For a few years he also freelanced in the then Cape Town Municipal Orchestra as French horn player. Until his late eighties he continued coaching students for their written music examinations.

In 1970 he gave a set of nine lectures on mysticism at the University of Cape Town's Summer School, and conducted several study groups on mysticism, Sanskrit, Hebrew and related subjects.

His first publication in the field of psychical experience was on angelic choirs in *The Hibbert Journal* of 1954 (reprinted in volume 3 of *Old and New Evidence on the Meaning of Life,*[06]). It was a wide-ranging account of non-physical states in which music, mostly song or chanting, has been reported. Its importance was in "pointing to the unique source from which all wisdom and goodness springs." In 1956 he published his first paper in the *Proceedings of the Society for Psychical Research* on out-of-body experience. He had joined the Society for Psychical Research in 1953, and was elected an honorary life member in 1999. He also had close connections with the American SPR.

In the late 1960s he came to the attention of the newly-formed South African SPR, centred at the University of the Witwatersrand in Johannesburg, and was invited to give lectures there. He also took part in a conference, Parapsychology in South Africa, held at the university in 1973 (Poynton, 1975), and a second conference held in 1980. He was awarded the Society's prestigious Marius Valkhoff Medal, and was elected a vice-president of the Society.

Overseas visits included taking part in a colloquium on Quantum Theory and Beyond organised by T. Bastin and D. Bohm at Cambridge in 1968, and the conference, Quantum Physics and Parapsychology, organised by the Parapsychology Foundation in 1974 in Geneva.

It could be said that in South Africa he was most widely known in musical circles, although in Cape Town he had colleagues, friends and students from diverse and wide fields of interests. Deep concern arose from what he saw as a miscarriage of justice due to failure of the judiciary to understand upheavals connected with spiritual development. Two cases in particular, of adolescent girls charged with murder, prompted an engagement with legal processes and, largely, the writing of volume 2 of his *Old and New Evidence on the Meaning of Life*.[00]

These activities and involvements made him a well-known figure in the Cape, and a special correspondent to the *Cape Times* wrote in November 2006, "Michael Whiteman, mathematician, musician and mystic, turned 100 last week, but is far too busy with his music and his writing to think of dying." In conversation he certainly gave the impression of someone half his age, an impression supported by a perfectly steady hand and dexterity at the piano. Yet it all came to a sudden end on the 5th of February, 2007, when he was found dead in his home by his daughter Sibyl, having suffered a heart attack. His wife Sona had died in 1982.

A website has been set up: www.whiteman.co.za.

REFERENCES

A select bibliography of Whiteman's publications in mysticism and psychical research was published in the *Journal of the Society for Psychical Research* 61(845) 1996: 254-255, and 71(886) 2007: 44.

Beauregard, M., Schwartz, G.E., & Miller, L. (2014). Manifesto for a post-materialist science. *Network Review* 115: 4-5.

Bohm, D. (1980). *Wholeness and the Implicate Order.* London: Routledge & Kegan Paul.

Boswell, J. (1791, 1992). *Life of Samuel Johnson.* London: Everyman.

Broderick, D. & Goertzel, B. (eds) (2015). *Evidence for Psi: Thirteen Empirical Research Reports.* Jefferson, N.C.: McFarland & Company.

Carr, B. (2015). Hyperspatial models of matter and mind. In Kelly, E.F. *et al., Beyond Physicalism: Toward Reconciliation of Science and Spirituality,* pp. 227-273. Lanham, Maryland: Rowan & Littlefield.

Crabtree, A. (2007). Automatism and secondary centres of counsciouness. In Kelly, E.F. *et al., Irreducible Mind: Toward a Psychology for the 21st Century,* pp. 301-365. Lanham, Maryland: Rowman & Littlefield.

De Pablos, F. (2013). *Brain at night: Scientific Foundations of Precognitive Dreams.* Murcia:Diego Marin Librero.

Eliot, T.S. (1969. *Four Quartets.* London: Faber & Faber.

Evans, A.J. (1921-1935). *The Palace of Minos at Knossos 1-1V.* London: Macmillan.

Gebelein, R.S. (2013). Physicalism. *Journal of Parapsychology* 77(2): 159-164.

Govinda, A. (1960). *Foundations of Tibetan Mysticism.* London: Rider & Co.

Gurney, E., Myers, F.W.H., & Podmore, F. (1886). *Phantasms of the Living* (2 vols). London: Trübner.

Heisenberg, W. (1959). *Physics and Philosophy: The Revolution in Modern Science.* London: George Allen & Unwin.

Honorton, C. (1975). "Error Some Place!" *Journal of Communication* 25: 103-116.

Husserl, E. (1931). *Ideas: General Introduction to Pure Phenomenology.* Trs. W.R. Boyce Gibson. London: George Allen & Unwin.

Jablonski, N. (2012). The struggle to overcome racism. *New Scientist* 215(2880): 26-29.

James, W. (1902). *The Varieties of Religious Experience.* London: Longmans, Green & Co.

Johnson, G.R. (2002). *Kant on Swedenborg: Dreams of a Spirit-Seer and Other Writings.* West Chester, Penn.: Swedenborg Foundation.

Jung, C.G. (1964). *Man and his symbols*. London: Aldus Books Ltd.

—. (1963). *Memories, Dreams, Reflections.* Trs. R. & C. Winston. London: Collins and Routledge & Kegan Paul

Kant, I. (1766, 2002). *Dreams of a Spirit-Seer.* In Johnson G.R. (ed.), *Kant on Swedenborg: Dreams of a Spirit-Seer and Other Writings*. West Chester, Penn.: Swedenborg Foundation.

—. (1781, 1934). *Critique of Pure Reason.* Trs. J.M.D. Mieklejohn. New York: Everyman's Library.

Kelly, E.F. (2007). Mystical experience. In Kelly, E.F. *et al., Irreducible Mind: Toward a Psychology for the 21st Century,* pp. 495-575. Lanham, Maryland: Rowman & Littlefield.

Kelly, E.F., Kelly, E.W., Crabtree, A., Gauld, A., Grosso, M., & Greyson, B. (2007). *Irreducible Mind: Toward a Psychology for the 21st Century.* Lanham, Maryland: Rowman & Littlefield.

Kelly, E.F., Crabtree, A., Marshall, P. (2015). *Beyond Physicalism: Toward Reconciliation of Science and Spirituality.* Lanham, Maryland: Rowan & Littlefield.

Marshall, P. (2011). The psychical and the mystical: boundaries, connections, common origins. *Journal of the Society for Psychical Research* 75: 1-13.

—. (2015). Mystical experiences as windows on reality. In Kelly, E.F., *Beyond Physicalism: Toward Reconciliation of Science and Spirituality,* pp. 39-76. Lanham, Maryland: Rowan & Littlefield.

Maslow, A.H. (1966). *The Psychology of Science: A Reconnaissance.* New York: Harper & Row.

Muldoon, S. & Carrington, H. (1929, 1973). *The Projection of the Astral Body.* New York: Samuel Weisner.

Myers, F.W.H. (1903). *Human Personality and its Survival of Bodily Death* (2 vols). London: Longmans, Green.

Neppe, V. (2011). Models of the out of body experience: a new multi-etiological phenomenological approach. *NeuroQuantology* 9: 72-83.

Nicholls, G. (2012). *Navigating the out-of-body experience: Radical New Techniques.* Woodbury, Minnesota: Llewellyn Publications.

Nilsson, M. (1927). *The Minoan-Mycenaean Religion.* Oxford: Oxford University Press.

Palmer, T. (2014). *The Science of Spirit Possession.* 2nd ed. Newcastle-upon-Tyne: Cambridge Scholars Publishing.

Popper, K.R. (1972). *The Logic of Scientific Discovery.* Revised edition. London: Hutchinson

Popper, K.R. & Eccles, J.C. (1977). *The Self and its Brain: An Argument for Interactionism.* Berlin: Springer-Verlag.

Poynton, J.C. (1975) (ed.). *Parapsychology in South Africa.* Johannesburg: South African Society for Psychical Research.

—. (1994). Making sense of psi: Whiteman's multilevel ontology. *Journal of the Society for Psychical Research* 59 (835): 401-412.

—. (2001). Challenges of out-of-body experience: does psychical research fully meet them? *Journal of the Society for Psychical Research* 65 (864): 194-206.

—. (2011). Many levels, many worlds and psi: a guide to the work of Michael Whiteman. *Proceedings of the Society for Psychical Research* 59 (222): 109-139.

—. (2011). Michael Whiteman: a biographical sketch. *Proceedings of the Society for Psychical Research* 59 (222): 141-143.

—. (2012). *Brief Authority.* Clovelly (Cape Town): Kima Global.

Ryder, K.C. (2010). *Heaven and Hell: Emanuel Swedenborg.* London: The Swedenborg Society.

Sidgwick, H. (1883). Presidential Address. *Proceedings of the Society for Psychical Research* 1: 7-12.

Smuts, J.C. (1926). *Holism and Evolution.* London: MacMillan.

Stevenson, I. (1997). *Reincarnation and Biology: Contribution to the Etiology of Birthmarks and Birth Defects* (two volumes). New York: Praeger.

Stoller, R.J. (1973). The sense of femaleness. In Miller. J.B., *Psychoalanysis and Women,* pp. 260-272. Harmondsworth: Penguin Books.

Swedenborg, E. (1758, 2010). *Heaven and Hell.* Trs Ryder, K.C. London: The Swedenborg Society.

Tart, C.T. (1975). *States of Consciousness.* New York: Dutton.

Tyrrell, G.N.M. (1953). *Apparitions.* Harmondsworth: Penguin Books.

Vitzthum, R.C. (1995). *Materialism: An Affirmative History and Definition.* New York: Prometheus Books.

Whiteman, J.H.M. (1961). *The Mystical Life: An Outline of its Nature and Teachings from the Evidence of Direct Experience.* London: Faber & Faber.

—. (1967). *Philosophy of Space and Time and the Inner Constitution of Nature: A Phenomenological Study.* London: George Allen & Unwin.

—. (1973). Quantum theory and parapsychology. *Journal of the American Society for Psychical Research* 67 (4): 341-360.

—. (1975a). Parapsychology as an analytic-deductive science. In Oteri L. (ed.) *Quantum Physics and Parapsychology* (pp. 181-204). New York: Parapsychology Foundation.

—. (1975b). The scientific evaluation of out-of-body experience. In Poynton, J.C. (ed.) *Parapsychology in South Africa* (pp. 95-108). Johannesburg: South African Society for Psychical Research. On-line Library, Society for Psychical Research.

—. (1975c). A three-tier ontology for parapsychology and modern physics. In Poynton, J.C. (ed.) *Parapsychology in South Africa* (pp. 124-134). Johannesburg: South African Society for Psychical Research. On-line Library, Society for Psychical Research.

—. 1977a. Parapsychology and physics. In Wolman, B.B. (ed.). *Handbook of Parapsychology* (pp. 730-756). New York: Van Nostrand Reinhold.

—. (1977b). The convergence of physics and psychology. In E. White & S. Krippner (eds) *Future* Science (pp. 289-308). New York: Doubleday.

—. (1985). Methods of "scientific mysticism" in the study of altered states in general. *Parapsychological Journal of South Africa* 6 (1): 1-33. On-line Library, Society for Psychical Research.

—. (1986a). *Old and New Evidence on the Meaning of Life. Volume 1. An Introduction to Scientific Mysticism.* Gerrards Cross: Colin Smythe.

—. (1986b). The mystical way and habitualization of mystical states. In Wolman, B.B. & Ullman M. (eds) *Handbook of States of Consciousness* (pp. 613-659). New York: Van Nostrand Reingold.

—. (1993). *Aphorisms on Spiritual Method: The "Yoga Sutras of Patanjali" in the Light of Mystical Experience.* Gerrards Cross: Colin Smythe.

—. (1994). Cognitive pathologies and the new paradigm in physics. *Journal of the American Society for Psychical Research* 88 (1): 81-86.

—. (2000). *Old and New Evidence on the Meaning of Life. Volme 2. The Dynamics of Spiritual Development.* Gerrards Cross: Colin Smythe.

—. (2003a). The problem of discarnate survival: towards a clarification of ideas in the light of a comprehensive life-world view. *Journal of the Society for Psychical Research,* 67 (871): 96-114.

—. (2003b). On the scientific validation and reality rating of reports of non-physial experience. *Journal of the Society for Psychical Research,* 67 (870): 37-51.

—. 2006. *Old and New Evidence on the Meaning of Life. Volume 3. Universal Theology and Life in Other Worlds.* Gerrards Cross: Colin Smythe.

INDEX

Abhiññā 72, 107, 115
Abrahamic religions 152
Acts of the Apostles 127
Action/actuation 43, 44, 45
Actualisation 5, 7, 8, 9, 12, 13, 17, 18, 19, 27, 31, 35, 36, 39, 42, 44, 45, 46, 47, 57, 95, 96, 112, 167
  and behaviour 57
  and existence 44
  differences according to state 9, 27, 31, 95
  in cycles 44, 45
  in thought-image sphere 17
  in time 5, 36, 39, 95, 96
  observer dependence 19, 48
  potentiality relationships 9, 13, 17, 35, 112
  subjective and objective factors 42, 44
Actual 36
Actuality 8, 9, 13, 13, 27, 45, 113
  non-physical 9, 31
  physical 8, 9, 31
Adolescence 57, 84
Adolescent breakdown 75, 78, 80, 81, 85, 87
Aegean civilisation 116
Agni 112, 113
Akarma 109, 110, 113
All-oneness, *kaivalya* 105, 108, 109, 110, 113
American Society for Psychical Research 177
An Introduction to Scientific Mysticism 2
Analysis, intelligible, phenomenological 4, 37, 106
Analytic-deductive, see under Science
Anatta 94, 95, 109, 114, 121, 134, 139, 144
Angela 58, 59, 60, 80, 89, 91
Angelic choirs 176
Angular momentum 171, 173
Anicca 115
Anima, see also female auxiliary sex 83
Anointed 125, 126, 127, 128
Anomia 128
Anti-Jewish 129
Aphorisms 104
Aphorisms on Spiritual Method 104, 109
Apocalypse, see also End of Days 125
Apparitions 13, 14, 15, 52, 53, 72, 73, 137
  Kant 52, 53
  non-physical nature 13
  of living 14, 72
Application 43, 76
Archetypal 139, 140, 141
  governance 2, 11
  humanity 105, 144
  light 127, 133, 134, 151
  man-woman 86
  reason 4, 100, 102, 139, 152
  structuring 101
  valuation 101
Aristotle 7
Aśoka 111
Assyrian advance 122
Astral travel 19
Atheism in Buddhism 144
Atta 94, 109, 114, 144
  compared with *ātma* 114

Ātma 114
compared with *atta* 114
Attachment 61, 75, 80, 94, 103, 107, 99, 100, 101, 105, 106, 107, 108, 109, 110, 114, 115, 135, 139, 140, 144, 162
Attitude 134
of incredulity 50
Authenticity 24, 27
Automaticity 43, 78
Auxiliary sex, see Sex
Avijjyā 78
Axe 109
Axioms 35, 36, 43

Bad, archetypal 140
Balfour, G. 62
Bar-Cochba rebellion 129
Barry 58, 59, 60, 80, 81, 89, 91
Bastin, T. 177
Beads on a string idea 67, 68
Beauregard, M. xvi, xviii, 5
Beautiful, archetypal 101,140
Becoming 44, 45, 78
Beholder as contributor 140
Bethlehem 127
Beyond Physicalism xvi
Bhagavad Gita 109, 110, 113
Bhava, Becoming 78
Bible
Authorised King James Version 100, 121, 123
New English 120, 123
Birth (foundational) centre 79, 80
Birth defects 62
Birthmarks 67, 68
Births, *jāti* 56, 107, 115
Bisexuality 82, 84, 118
Bishops, Diocesan College 176
Body
kinds of 141
non-physical 19, 22, 132
physical 15, 20, 22, 25, 132, 134
separated 20
spiritual 123
Bohm, D. 29, 177
Boswell, J. 8
Boundary conditions 112
Brahma-yoni 112
Brahaminism 111, 114
Brain 3
Brain lateralisation 85
Breadth, *y,* 42, 165
Breath control, *prāṇāyāma* 110, 111
Bṛhad-āraṇyaka Upanishad 45
Brief Authority 112
Broderick, D. & Goertzel, B. 53
Buddha, Gotama xiii, 27, 56, 63, 67, 96, 101, 104, 114, 120, 134, 148
Buddhism 6, 37, 38, 76, 78, 94, 96, 104, 111, 114, 140, 144
Buddhist canons, teaching 5, 48, 72, 96, 101, 105
Buddhist path 144

Cambridge University 154, 155, 175
Cambridge University Library xx
Cape Times 177
Cape Town 56, 58, 176, 177
Municipal Orchestra 176
University of xvi, 176
Carr, B. xxi, 18, 42
Carrington, H.155
Carroll, S. 175
Causation 51, 112
Centre, *chakra* 85, 86, 108, 118, 156
birth, *mūlādhāra* 85
foundational (birth) centre 86
head 86
heart 85, 86, 108
navel 85, 86, 108
noetic 86
solar 86, 108
substantial 86
supernal 86, 108
throat 85, 86
Chaitanya 83
Chakra, cakra, see Centre
Change 160
and time 160
Channelling 81

Checking, mind involvement 144
Child 122, 123
Childhood memories 60, 62
Christ, see also *Christos* 126, 127
Christian 139
Christian Era 114, 130
Christian mystics 120
Christianity 5, 78, 106, 120, 140, 142
Christians 129
Christos 126, 126, 127, 128
Clairvoyance 3, 17, 24
Classification 11, 15
Cloak of Ideas 4, 34, 37, 68, 100
Clock-time 11
Cognitive pathology xviii
Collectedness, *samādhi* 110
Co-mind, CM, Contributory mind 6, 14, 61, 62, 63, 64, 67 71, 72, 73, 74, 79, 88, 95, 109, 115, 121, 135
  higher 79, 81, 85
  lower 79, 80, 81, 85, 89, 90, 91
  objectification 88
Coming, Second 128
Coming-and-going of influences 56, 58, 61, 64, 66, 73, 74, 75, 115
Communication of thought 21
Complacency 44, 78
Conceptual analysis 33
Conditioning 78
Conduct disorder 135
Conflicts 148
Conscience 77
Consciousness 37, 40, 107, 139
  altered states 54, 161
  double 18
  line of 64, 65, 66, 67, 68, 69, 70, 73, 139, 151
  physically-based 11
  pure 37, 41
Contemplation, *dhyāna* 110
Contests
  inner 44, 74, 75, 77, 78, 79, 80, 82, 84, 85, 87, 91, 103, 123, 137, 143, 148, 149, 151, 157
  near-ultimate 79
  ultimate 79, 120, 125, 157
Contributory minds, co-minds, CMs 6, 61, 63, 64, 65, 66, 71, 82, 109, 121, 124, 129, 138, 143
  higher 78
  lower 78
Control centres, *chakras,* see Centre 85, 86, 108
Core identity 61, 63, 64, 65, 81, 83, 90, 102, 103, 115, 121, 123, 127, 129, 134, 135, 137, 139, 140, 143, 144, 161
  female 80, 83
  male 80
  pre-existence 65
Corinthians, letter 127
Corporality 78
Corporate structure, see under Personality , 103
Cosmic distance, interval, time 163
Counterfeit 43, 44, 76, 78, 79, 81, 84, 85, 139, 148, 160
Crabtree, A. 62
Crete 116
Crime 88
Critique of Pure Reason 51, 54
Cross 127
Cross dressing 84
Crucifixion 126, 127
Cycles,
  creative fourfold 43, 95, 103, 122, 127, 130
  five-fold 107, 108
  fixated 44
  free 44
  lower 44, 81
  objective 47, 48, 159, 160, 161
    combined with subjective 47, 160
  sixteen-fold 5, 6, 47, 48, 159, 160, 161, 165
    objective 6
    subjective 6, 165
  subjective 47, 48, 159, 160, 161, 167

combined with objective 47, 160
world 48, 161

Dactyls 116
Damascus 127
Dancer 149
Daughter, mystical meaning 83, 102, 121, 123, 124, 129, 132
Davies, P. 174
Day of Judgement, the Lord, see also End of Days, Last Judgement, Third day 137, 141, 142, 144
de Beauvoir, S. 88
de Pablos, F. 25
Death 158
physical 65, 140, 151, 158
Deed-attachment 144
Deformity 67, 68
Deletion of old self 142, 143
Demarcation/distinction
mystical/psychical 3, 27
mysticism/science 3
physical/non-physical 11, 12, 13, 14
Depersonalisation 80, 81, 88, 90, 91, 96
Derealisation 80, 81, 88, 91, 96
Description of experiences 156
Desire 144
Destiny 95, 96
Detachment 11, 30, 34, 38, 51, 106, 110
Development
spiritual 90, 97, 109, 135, 143, 148, 151
Dhamma 94, 109, 114, 121, 134,139
Dhammapada 114
Dharma 114
Dhyānas 37, 110
Differences in actualisation 31
Dīgha Nikāya 56, 61
Diodoros 116
Dirac, P.A.M. 161, 167, 157, 159, 161, 172, 174
Directions, see Functions
allegorical 133, 136
Disciples 129
Disposition 61, 64, 65, 74, 109, 148, 140, 151
affected by physical acts 137
Dissociation 13, 15, 16, 17, 18, 20, 22, 30, 37, 39, 73, 76, 79, 80, 81, 88, 89, 107, 135, 154, 156, 157, 168, 174
higher 38
lower 38, 39
Distance
cosmic 163
physical 73
Divine, divinity 5, 100, 127, 141
Divine Humanity 105, 106, 130
Divine Man 105, 108
Divine Source 72, 73, 76, 77, 80, 96, 97, 103, 120
Dogmatism 52
Double 14, 72, 73
mimicking someone else 73
Dreams, dreaming 25, 160
lucid 25
telepathic 25
Dreams of a Spirit-Seer 52, 53, 54
Drug influence 38
Dualism 8, 141
Duality 8
Dukkha 114, 115, 144
Duplicate physical 13, 31, 133, 134,
Dynamics of Spiritual Development 101

Earths, three Vedic 135
East
in four-fold cycle 43, 44, 45, 111, 116, 117, 118, 122
in paranoia 136
in sacred marriage 136
Eavestaff, D. 176
Eccles, J.C. 150
Eckhart, J. 78, 83, 141
Eddington, A.S. 172
Einstein, A. 34, 159, 165

Einstein-Bose statistics 171
Electromagnetic field equations 167
Electron spin 171
Elementary particles 12, 29, 48, 167, 169, 171, 173, 174
  physical/non-physical question 12, 29, 48, 167, 168
Eliot T.S. 38, 94, 149
Embodiment, physical 66
Empathy 115
Empiricism 2
Enclosure 8, 94, 140
Ends 44, 45, 95
End of days, see also Last judgement, Third day, Day of the Lord 120, 135, 141, 142
End-of-the-world quality 79
Engagement in life 149
Enlightenment 2, 6, 50, 51, 54
Entanglement, quantum 12
Epiphany 150
Epistemology 12
Epistles 126, 128
Epoché 37, 38, 46, 105, 162
Erfahrung 51
Error Some Place 2
Escape 79, 80
  dissociated 74
  psychotic 74
ESP 2
Essence 5, 35, 36, 44, 45, 46, 86, 95, 122, 154
  essential insight 35, 36, 154
Ethical viewpoint 90
Euclid 39
Evans, A.J. 116
Evidence for separation, OBE 25, 27
Evil 80
Existence 5, 29, 35, 36, 44, 45, 86, 112, 122
Experience 52
  ante-world 133, 134
  doubly-conscious 20
  mystical 8, 9,15, 25, 27, 30, 33, 39, 121, 140, 159
    description 133, 156
  near-death 16, 30
  non-physical, other-world 2, 3, 4, 32, 39, 51, 134, 160
  out-of-body, see also separative 8, 15, 16, 19, 20, 27, 160, 177
    spatial relationships 16
  paranormal 17
  personal 68
  physical 4, 22, 27, 28, 51, 148, 160
  pre-mystical 27
  psychical 9, 27
  separative 15, 16, 19, 22, 23, 24, 25, 30, 31, 32, 132, 134, 140, 155
    indubitable 27
    limitations on recollection 138
    reality of 24
    spatial relationships 16
    veridicality 25, 27
Extrasensory perception 25

Facial expression 35
Faith 76, 77, 78, 96, 102, 106, 116, 117, 136, 139, 143, 156
  pistis 125
False, archetypal 130
Fantasy 32
Fashion 121
Father, mystical meaning 127, 129, 130,140
Female/feminine functions (S, N) 77, 81, 84, 85, 86, 127
Female
  auxiliary sex in males 83, 84, 86
  core identity 83, 86
  qualities 119
Fields 12, 167, 170, 174
Fifth upward element 108, 109
Fire, mystical meaning, see also *agni* 137, 156
Five-fold pattern, cycle, see Cycle 116, 108

Fixation 29, 31, 36, 40, 51, 75, 80, 81, 94, 140
Flesh 128, 129
Florence 65
Forehead 86
Foreigners, mystical meaning 102, 123
Four directions 40
Four Quartets 33, 38
Four-fold cycle , see Cycle
Fourier analysis 170
Freedom 90
 from physical space and time 39
Free will 18, 39, 90
Freud, S. 43, 62, 75, 83
Functions, creative 43, 44, 46, 75, 76, 77, 84, 85, 87, 111, 112, 113, 116, 117, 122, 135, 136
 auxiliary 86
 feminine (2 & 4) 77, 84, 87, 117, 124, 127
 masculine (1 & 3) 77, 84, 87, 117, 124, 127
 principal 86
 time-like 35
Fusion 110

Galatians, letter 127
Galileo 42, 46, 50, 160, 161
Gebelein, R.S. 1
General Index of Reality 25, 26, 27, 44, 71
Ghosts 52
God 5, 78, 86, 100, 102, 106, 112, 124, 127, 128, 129, 139, 141, 143,
 buddhism, no reference 144
 decline in Whiteman's use 93
 dice-playing 151
 ideas in 156
 unsuitable use 152
Godhead 12, 139, 141
God-image 88
God-man, *Puruṣa* 141
God-source 112
Gonville & Caius College 175
Good , see also Right 76, 125, 128, 140
 archetypal 101, 140
 idea of 156
 potential for 158
Gospels 126, 128
 synoptic 126
Govinda, A. 56
Gradient operator 166, 167
Grant, J. 71
Grasping 141
Gravity 12
Great Mother 69, 86, 117
Great Stride of Vishnu 45, 112, 113
Greece, classical 50
Greek 5, 100, 126, 127, 128, 129
Grounding 44, 45
Group theory 163
Guidance 3, 156
Guilt 73, 89, 90, 92, 93
Gurney, E. 71

Hallucination 24, 31
Hamartia 92, 93, 125, 127
Hamilton, W.R. 169
Hamlet 95
Handbook if States of Consciousness 75
Harmonics 168, 170, 171
Head, forehead centre 80
Heart centre 79, 80, 99
Heaven and Hell 54, 60
Heavens 29, 142
Heavens, three Vedic 135
Hebrew 5, 100, 129
Height, *z*, 42, 165
Heisenberg, W. 7, 8, 12, 29, 94
Henin 70, 71
Hierarchical/hierarchy 12, 46
Higher Criticism 126, 128
Highgate School 175
Hīna 101
Hīnayāna 101
Hindley, M. 88
Hindu 114
Hinduism 114

Holism and Evolution xviii, 51
Honorton, C. 2
Hosea 143, 144
Hostility to mysticism and psychical research xvii, xviii, 2
House of personality 56, 61, 79, 115
Human condition 162
Humanity, archetypal, universal, divine 106, 112, 130, 141, 142, 144, 160
Husserl, E. 4, 34, 36, 37, 40, 41, 46, 68, 83, 100, 105, 106, 107, 113, 151, 162
Hymn to Agni 112
Hyperspace 42
Hypostases 44, 45
Hypotheses 37
  physicalistic 162
Hypothetical-deductive, see under Science

I AM 139, 140, 141
Iaso 126
Ideas 37
  cloak. See Cloak of
  life and truth 133
  pure 37
  structure of 34
Identification 61, 70, 140
Identity 160
  mystical 121
  sexual 77
Ieso 126
Iēsous 126, 128, 129
Immediate seeing 35, 106
In the world but not of it 149, 157
Incarnation
  loose 60
  strict 60, 61
Incredulity 50, 53, 54
Index of Reality, see General Index of Reality
Indifference 149
Individual 8, 75
  in charge, IC 61, 62, 63, 64, 65, 66, 67, 72, 73, 80, 81, 82, 88, 89, 95, 110, 135
  life and physical laws 161
Indivisibility of the phenomenon 8
Infection, psychic 73
Influence, psychic
  benign 73
  malign 73
  neutral 73
Inner constitution of nature 2, 4, 48
Inner teaching 157
Inner voice 156
Insight 31
Insistent directive 58, 73, 81, 89, 90, 91
Intelligibility 152
Intelligible structure 34
Intelligible time 32, 33, 34, 36 38
Interval, cosmic 163
Involvement checking 144
Iron rule xiv, 51, 53, 54
Irreducible mind xvi
Īsa Upanishad 108
Isaiah 48, 83, 120, 121, 121, 122, 123, 129, 135, 142, 143
Islam 103
Īśvara 105
Israel, mystical meaning 132, 133, 143

Jablonski, N. 52
Jackie 72
Jacob 120, 143
James, W. 62, 76
Jāti 56, 107, 115, 144
Jerusalem, mystical meaning 120, 122, 142
Jesus, see also *Iēsous* 122, 126, 127, 129, 130, 139, 142
Jews 129
Jhānas 33, 38, 41, 94, 106, 107, 156
John the Baptist 93
Johnson, G. 52, 53
Johnson, S. 8
Joshua 126
Judah, mystical meaning 120, 142
Judas Iscariot 120

Jung, C.G. 43, 62, 83, 84, 162
Juridical viewpoint 90
Justice 90

Kaivalya 105, 108, 109, 113, 156
Kalās 6, 45, 48, 111
Kāma 78, 144
Kant, I. xvii, 13, 30, 51, 52, 54, 151, 162
Karma 66, 69, 109, 110, 113, 144
Kaṭha Upanishad 66
Kauṣītakī-Brāhmana Upanishad 63
Kelly, E.E. xvi, xvii, xxi, 13, 19, 42, 50, 53, 61
Kevala 108
Kidney 16
King
  mystical meaning 121, 124
  worldly 123
Knossos, Palace of 116
Know Thyself 62
Krippner, S. xxi
Kundalini Yoga 87

Lamb 122
Last judgement, last times 137, 143
Law, divine 140
Left, direction 78
Length, *x,* 42, 165
Lenz, H. 57
Letter 100, 103
Letter vs spirit 100, 102
Levi-Civita, T. 151
Leviathan 125
Levy, R. 71
Lichtigfeld, A. 108
Life
  continuation 151
  light of 133
  meaning of 137, 138, 140, 142
  meaning to 138
  otherworldly 132, 133
  physical 132
  purpose of 138
  source 123
  termination 140
  unitive 108
Life-plan 95
Light
  archetypal 133, 134, 151
  divine 135, 136
  mystical meaning 121
Lighting effects 31
Limbs, non-physical 132
Limbs of enlightenment 110
Limbs of Yoga 110
Line of consciousness, see under Consciousness 55
Logos 129, 141
Lord 122, 123, 126, 127, 135, 142, 143
Louis XVI 95

Mahāyāna 101
Major Possessive Disruption 64
Male/female differences 81, 82, 86, 119
Male/masculine functions (E , W) 77, 81, 84, 85, 86, 127
Malice 109
Man 127
Manifestation 44, 45, 112
Map of time 39
Marius Valkhoff Medal 177
Marlene 57, 58, 60, 73, 80, 84, 88, 89, 91
Marshall, P. xvii, xxi, 3, 38
Maslow, A.H. xviii
Materialism xvi, 2, 5, 50, 129, 130, 149, 150, 151, 152
  promissory 150
Materialists 44, 128, 150
Mathematics 161
  Department of, U.C.T.
    Applied 176
    Pure 176
Matter 42, 140, 141
Matterism xvi
Maurya Empire 111
Māyā 8, 9, 29, 94, 140, 150
Maxwell, J.C. 161, 167
McDougall, W. 62

Me vs I 139
Meaning of life 148, 149, 150, 151, 153
Meaning in life 151
Meaning to life 149
Means 43, 44, 45, 95, 116
  universal 89
Measurement 51, 163
Mechanics 35
Mechanistic regime xviii, 51, 53
Mediate identification, see also Identification 70, 72, 108
  of the past 70, 71
  of the present 72
Meditation 41
Mediumship 81
Memory 25, 68, 70
  inner 151
Mendelssohn, M. 53
Mens rea 92
Messiah 142
Messiah, A. 171
Metanoia 92, 93, 125
Mind 3, 8, 42, 140
  contributory, see Co-mind, CM
  irreducible xvi
  universal 18
Minoan culture 3, 68, 71, 116-119
Mistaken, archetypal 140
Mistranslations, see Translations
Mokṣa 95
Moon 33, 34
Moors Murders 88
Morality 16
Motivation 109, 134
Mountain of the Lord 142
Mozart, W.A. 38
Mūlādhāra, see also birth centre 85
Muldoon, S. 155
Multiple frequency 22
Multiple personality disorder 64
Multiple space/level 4, 22, 57, 157
Multiplicity 78
Muṇḍaka Upanishad 112, 113
Murder 56, 91, 92, 177
Music , 38, 154, 157
  Whiteman's involvement 161
Musical
  melodies 108
  sound 170
Myers, F. 62
My self 139
Mysteries, Cretan 116
Mystical
  definition 101
  derivation of physical law 163
  development 3
  distinction from psychical 3
  experience, see Experience
  initiation 127
  life 154, 158
  living the 125
  research xviii
  state 14, 25
  teaching 144, 148
  view 2, 90
  way 37, 130, 144, 152, 156
  world-view 148, 152
Mystical Form Liberation 32
Mystical transformation 73, 75, 77, 90
Mystical Way 37, 130, 144, 152, 156
  guidance 145
Mysticism xvii, xviii, xix, 2, 3, 11, 54, 152
  and intelligibility 152
  and physics 147, 148
  and psychical research 3
  and reason 152
  and religion 152
  and science 2, 3
  and theology 152
  definition 2, 3, 11, 19, 34, 96, 101, 106
  negative view of 2, 151
  science of 54
  scientific 2, 3, 4, 5, 54
  Tibetan 56
Mystics 76, 151
Mystisch 151

Nancy 20, 134
National Museum of Denmark, Copenhagen 117, 118
Nations, mystical meaning 120
Natural attitude 37
Naturalism, one-level 4, 6, 13, 17, 18, 30, 31, 42, 52, 73, 127
  definition 4
Nature 150
  inner constitution 3, 5
Navel (solar) centre 79, 80, 99
Nazareth 127
Negligence 90, 91, 92
Neo-Platinism 140
Neppe, V. xxi, 19
Neurosis 16, 135
New Testament 80, 103, 148
Newton, I, 34, 50, 169
Nibbana 144
Nicca 115
Nicholls, G. 19
Nilsson 108
Nirmāṇa-cittāni 109
Nirodha 107
Noetic cycle 86
Non-attachment 43, 61, 76, 105, 106, 128, 140, 149
Non-local, non-locality 11, 12, 17, 40, 73, 169
Non-physical 2, 3, 4, 11, 73
  definition 4, 11
  different from physical 11
Non-physical beings 154
Non-physical state 4, 15, 63
  limited prospect of spiritual growth 137
Non-speculative state 106
North, in four-fold cycle 43, 44, 45, 117, 118, 122
Nostradamus 95
Number systems 43, 45, 46

Obedience 76, 77, 78, 79, 80, 81, 82, 86, 105, 106, 116, 133, 136, 139, 144, 149, 150, 154, 155, 156, 158
  in females 77
  in males 77, 86
Objectification 63, 64, 76, 135
Objective factors 8, 42
Objective non-physical experience 51
Objective space-time and subjective functions 47
Objectivity 160
  and space 160
Objects of attention 160
Observation 12, 52
  occasion of 5, 7
Observer 133, 164, 174
  "gas" 163, 164
  state and actualisation 26
Old and New Evidence on the Meaning of Life 3, 75, 177
Old Testament 80, 102, 106, 120, 122, 129, 135, 141
One-level idea, see Naturalism
Opening 13, 15, 16, 73
Operating theatre 30, 31
Opposites 78
Other-worldliness 2
Out-of-body detection 16
Out-of-body experience, see Experience, out of body
  research pitfalls 31
Over-knowledge, see *abhiññā*

Pali xiii, 5, 100, 101, 114
  cf. Sanskrit 92, 101, 114
Palmer, T. 62
Paranoia 135
Parapsychologists 2
Parapsychology 2 ,4, 27, 151
Parapsychology Foundation 177
Parapsychology in South Africa 177
Parousa, transformation 128
Particles, see under Elementary particles, and Physics
Pascal, B. 108, 156
Patanjali 104
Pelman course 41, 155

People, peopling, mystical meaning 102, 120, 121, 123,141, 142, 143
Perception
  mode of 54
  mystical 8
  non-physical, psychical 8, 31
  physical 8, 31
Persona 56
Personality 4, 57, 62, 64, 66, 67, 75, 79, 102, 160
  attachments 144
  changes 51, 149
  corporate structure 6, 60, 61, 62, 63, 64, 66, 67, 73, 82, 84, 102, 109, 120, 121,141
  definition 64
  development 122
  generation , see *jāti* 109
  house of 56, 57, 92, 93, 94
  multiple 56, 64
  phases, see *jāti* 56, 64, 107, 115
  physical/non-physical appearance 83, 123, 132
Perspectives 7
Petticoats 68, 69
Phenomenology 4, 5, 34, 35, 37, 51, 68, 100, 107, 113, 151, 154, 162
Philosophy of Space and Time 4, 48
Physical 12, 48, 148
  appearance 83
  acts, affecting disposition 148
  corroboration 21
  definition 11, 30
  different from non-physical 12, 13
  distance 67
  duplicate 12, 27, 124
  laws 30, 163
    and individual life
  life, importance for spiritual progress, development 144, 148, 149
  normal 4
Physicalism xvi
Physicists 12
Physics xviii, 12, 34, 35, 42, 43, 45, 46
  and mysticism 159
  classical xviii, 42
  field laws 161
  mathematical xvi, 3
  particle 6, 12, 29, 48
  union/link with psychology 6, 8, 36, 42, 43, 46, 47
Physics and Philsophy 7
Pipher, M. 84
Pistis 125
Plan 95, 96
  divine 96
  life 95
  world 95, 96, 97
Planes 29
  lower 158
Plato 48, 148, 150
Pleasure principle (Freud) 75
Poincaré, 43
Popper, K. 33, 150
Possession 62, 73, 80, 81, 89, 91, 92, 135
Post-hypnotic suggestion 16
Potentia 7, 12
Potentiality, see also Substructure 5, 7, 8, 9, 12, 17, 18, 19, 27, 29, 31, 35, 36, 39, 44, 45, 48, 57, 64, 83, 95 111, 112, 141, 161, 167
  essence 36, 44
  fields, sphere 9, 17, 18, 27, 31, 35, 39
  for good 158
  impersonal 9, 42
  individual 9
  mental 9
  non-physical 9
  personality 57
  objective 9
  related to actualisation 9, 17, 35, 112
  subjective, personal 9, 42, 57
  time-ordered 8, 39, 95
Potentiality fields 9, 27, 31

time-ordered 8, 39, 95
Power, transcendent 140
Powers, voluntary 78
Poynton, J.C. xvi, 24, 31, 177
Prāṇāyāma 110
Precognition 17, 18, 39, 46, 51, 52, 96, 157
Prejudice 52
Presentiment 17
Primary qualities, see Qualities
Primordial concept 35
Principal sex 78, 80
Principle of Sufficient Reason 65, 151
Projection, power of 8, 29, 140
Propaganda 25
Psalms 123, 124
Pseudoscience xviii, 53
Psi see psychical
and mysticism 3, 4, 6
psi and quantum theory 12
psi phenomena/research and
psi performance 76
Psychiatric opinion 88, 89
Psychic ability 52, 59
and observer 19
Psychic attack 59
Psychic infection 73, 91
Psychic space 30
Psychical/psi phenomena xviii, 3, 4, 8, 10, 11, 12, 30, 47, 129
and quantum theory 12
and substructure
distinction from mystical 3
Psychical research xvi, xvii, xviii, xix, 2, 3, 5, 17, 33, 34, 54
and mysticism 3, 27, 28, 62
and science 3
evidence 51
fallacious 2, 52, 53, 54
Psychopathology 80
Psychophore 68
Psychokinesis 17
Psychology 8, 42, 43, 46
link/union with physics 42, 43, 46
Psychometrist 71
Psychosis 16, 135
Puberty 87
Punishment 88, 90, 92, 93
Purpose 43, 44, 45, 95, 96, 112, 116
Divine 123
in life 151
of life 138, 142, 149, 153
for all humanity 95
universal 89, 90
Purposeless 149
Puruṣa 105, 106, 108, 113, 124, 126, 141
Pythagoras 163

Qualities
primary 42, 43, 46, 160, 161
secondary 42, 43, 45, 46, 161
Quanta 12
Quantum
field theory 48, 159, 161, 170
physics 112
theory 7, 8, 12, 43, 148, 161
Quantum Theory and Beyond 177
Quantum Physics and Parapsychology 177
Quasi-spatial concepts 42

Race 52
Racism 52
Real being 32, 139, 153
Reality 7, 8, 9, 20, 22, 23, 24, 36, 101
index, see General Index of Reality
non-physical 4
of separative experience 24ff
physical 4
test of 24
Reality impression, range 22
Reality principle (Freud) 75
Reason 2, 52, 129, 154
archetypal 4, 100, 102, 129, 139, 140, 141, 152
boundaries and limits set by reason 53

divine 140
Reasoning 2, 154
archetypal 4, 100, 102, 129, 139
Recklessness 90, 91, 92
Recollection 40, 76, 86, 87, 100, 102, 106, 107, 108, 149, 154, 155
active 40, 41, 44, 76, 78, 79, 85, 106, 107, 117, 135, 155
continuous 41, 76, 77, 78, 79, 85, 86, 105, 107, 117, 135, 155
in females 80, 86
in males 80, 86
different powers in physical/non-physical life 138
practices 154
Reed 27
Reflection 38
Reincarnation 65, 66, 67, 68, 69, 70, 71, 72
and children 66, 67, 68
loose 66, 67
strict 66, 67
Release 36, 37, 40, 41, 43, 51, 57, 61, 63, 80, 93, 94, 95, 96, 102, 103, 107, 108, 109, 132, 135, 140, 141, 143, 144, 149
Reliability 24, 27
Religion 5, 118, 119
and mysticism 152
Religious beliefs, thinking, philosophies 2, 5, 102, 118, 129, 140, 152
Hebrew 120
Religious practice 2
Remnant of personality/Israel 142, 143
Remote viewing 17, 24
Renaissance xvi, 54
Repentance, see *metanoia*
Research, see Psychical
Resonance 17
Responsibility 73, 89, 90
Retrocognition 68, 70, 71, 115
Rhodes University 176
Ricci, G. 151
Rig Veda 62, 100, 104, 105, 112
Right and Good 2, 76, 77, 82, 83, 93, 102, 123, 126, 142, 144
Right, direction 78
Robertson, H.P. 163
Rogo, S. 12, 29
Rome, Romans 129, 141
Ṛta 140
Ruling identity, RI, 64, 121, 142

Sacred marriage 69, 116, 117, 124, 136
Saint Bernard 83
Saint James 117
Saint John 117, 119, 120, 128, 130
Saint John's Gospel 126, 128, 129, 130, 140
Saint John of the Cross 83, 154
Saint Mark 126
Saint Matthew 125
Saint Paul 83, 85, 100, 125, 126, 127, 128, 130, 141
Saint Peter 127
Samādhi 107, 110
Saṁsāra 56, 93, 94
Saṃyama 107
Saṃyoga 110
Sāṅkhya philosophy 104, 105, 111, 114
Sāṅkhyan Era 114, 130
Sanskrit 5, 100, 104, 105
vs Pali 101, 105, 114
Sarx 129
Satanism 59, 81, 89
Sati 98
Saviour, Anointed 126, 127, 128
Scandal 50
Schizophrenia 135
Science, xvi, 2, 11
analytic-deductive 33, 34, 35, 47
classical xviii, 152
hypothetical-deductive 33, 34, 68
of mysticism and psi xviii, 3, 130
post-materialist xvi, xviii, 5
Science and spirituality xvi, 42

Science, mysticism and psychical research xix, 2, 3, 6, 33, 105
Scientific thinking 33
Scientific mysticism 2, 4, 5, 33, 37, 52, 102, 145, 149
Scientific viewpoint 90
Scientifically usable 24
Scientists 29
Scissors murder 58, 88
Script 90, 95
Second Coming 128
Secondary qualities, see Qualities
Sedenion algebra 46, 161, 165, 171
Self 75, 94, 103, 109, 114, 139, 144
  higher 114, 115
  ordinary 32
  real 32
  very 156
Selfless, selflessness 77, 78, 84, 94, 114, 119,134, 139, 140, 144
Selfless real being 32, 139
Self-beautification 121
Self-centredness 109
Self-discipline 106
Self-evident truth 106
Self-satisfaction 43
Self-will 43, 78
Separation (out-of-body experience) 15, 16, 29, 63, 65, 72, 73, 81, 107, 108, 133, 134, 140, 148, 149, 151, 154, 155
  appearance in 83, 134
  fantasy 23
  primary 19, 20, 25, 30, 148
  secondary 20, 21, 70, 72, 73, 134
  tertiary 20, 21, 22, 83, 108, 138
Separative experience, see Experience, separative
  incipient 143
  research pitfalls 27
Septuagint 126
Sex
  auxiliary 83, 84, 87, 118, 124
  core 82
  principal 84
Sexuality 82, 85
Shamanic phase/initiation 80, 157
Sibyl 176, 177
Sidgwick, E. 21
Sidgwick, H. 50, 53, 54, 101
Signet ring, Minoan 108
Sin , see *hamartia*
Six-dimensional system 46
Sixteen-fold system/cycles, see under Cycles
  structure of concepts 47
  physics 46, 47
  psychology 46, 47
Skills 4, 16, 40, 43, 75, 76, 78, 79, 86, 89, 90, 92, 95, 96, 103, 117, 143, 148, 149
  of intelligible time 38, 40, 75, 86, 90
Smear-words 92
Smuts, J.C. xviii, 51, 53
Society for Psychical Research xvii, 50, 177
  Archives xx
Solar centre 80
Son, mystical meaning 83, 102, 103, 121, 122, 123, 124, 127, 129, 130, 132, 144
Son of Man 125
Sona 176, 177
Soul 53, 109
Sound wave 174
Source, 124, 126, 140, 141, 146
  divine 78, 79, 82, 87, 105, 106, 112, 121, 125, 129, 134, 135, 136, 142, 144, 153, 155, 158
  goodness and Wisdom 102, 105,
  higher values 72
  life and truth 133
  right and good 2, 102, 120, 127, 129
South
  in four-fold cycle 43, 44, 45, 116, 117, 118, 122
  in sacred marriage 136
  schizoid 136
South Africa iv, 176

South African Society for Psychical Research xxi, 177
South African Society of Music Teachers 176
Space 4, 34, 39, 160
  breadth 39
  dimensions 4, 39, 46
  duplicate 30
  height 39
  intelligible 22
  length 39
  multiple 4
  non-physical, other, psychical, spiritual 4, 13, 15, 16, 17, 25, 26, 29, 30, 39, 61, 160
  physical 4, 15, 16, 30, 39, 160
  rational structure 34
  substructural 163, 168
Spaces, interpenetrating 22
Space-time
  components *x, y, z, t* 46, 170
  measurement 151
  substructural 168
Spatial dimensions, variables *x, y, z* 46, 165, 168, 170
Speculation 2
Spinor 171, 173
Spirit, universal 140,141
Spirit/s 52, 60, 62, 128, 129
  conjunction of 60
Spirit vs letter 100, 102
Spirit Release Therapy 62
Spiritual growth/development 90, 143
  in non-physical worlds 137
  in physical world 134, 137, 138
Spirituality xvi, 89
Staffords School 176
State, non-physical 15, 63
  of observer 31
Stevenson, I. 4, 29, 67, 68
Stoller, R.J. 61
Stoppage of time, epoché 35, 37, 38, 40, 105, 106, 107, 155, 162
Stream-winning 144
Stress 75, 79, 103, 143, 144, 156
Struggle 141
Subfunctions 45, 46, 122
Subjective factors 8, 42
Substantial cycle 80
Substructure, see also Potentiality 5, 6, 7, 57, 64, 155, 159, 160, 161, 163, 168, 169
  in mysticism and physics 159
  non-physical 159
Suffering 114, 127, 144
Suicide 88
Sun
  divine 158
  physical 33, 34
Supernal centre 80, 99
Superposition 174
Sutras 104, 105, 107
  style 95
Swedenborg, E. xvii, 52, 53, 54, 60, 61, 62, 67, 83, 127, 128, 132, 133, 136, 148

Taints 109
Telepathy 3, 17, 51, 115
Tensor analysis, calculus 46, 159, 160, 163, 165
The Foundations of Mathematics 176
The Importance of being Earnest 2
The Mystical Life 3, 11, 101, 102, 154
The Projection of the Astral Body 155
The Yoga Sutras of Patanjali 105
Theological beliefs 102
Theological viewpoint 90
Theology 5, 100, 139, 152
  and mysticism 152
  universal 102, 112, 114, 126, 152
Theoretical interpretation 102
Theory 2
  of relativity 159
Theos 5, 100, 139
Theosophy 67
Thessalonians 128

Thinking
  conventional 6, 141, 154
  scientific 33
Third day, see also End of days, Last judgement 134
Thought-image sphere 12, 17, 39
Thumb 107, 108
Time 3, 4, 34, 37, 38, 39, 46, 78, 155, 160
  become like space 38, 155
  clock-time 11, 38
  cosmic 163
  dimensions, see Time-like dimensions
  intelligible 38, 40, 75, 86, 90
  levels 39
  non-physical 35
  rational structure 34
  stoppage 30, 31, 36
Time-like dimensions, variables, functions
  τ, actualisation time 39, 46, 161, 165
  t, clock-time 11, 39, 46, 161, 165, 168, 170
  T, intelligible, potential, cosmic time 39, 46, 161, 165
Timeless recall 155
Time-ordered potentiality 17, 18, 39
Time-release, timelessness 38, 39, 40
Transcendence 25
Transformation
  personal 3, 14, 21, 96, 102, 128, 132, 134, 135,142, 144, 153
  measurements160, 164, 165
  mystical 83, 81, 160
  unitive 16, 123, 124
Translations 100, 101, 102, 104, 114, 120, 125, 127, 128
Transvestism 84
Tree of Life 127
Trick, see *māyā*
True, archetypal, see also Good and True101, 125, 140
Truth 37, 162
Tyrrell, G.N.M. 13, 72

Ugly, archetypal 140
Ulman , M. 75
Ultimates 3
Unconditioned 77, 110
Union, transforming 114
Unitive life 108
Unitive transformation, state 16, 32, 96, 108, 110, 144
Unity, transcendent 140
Universal reason 34
Universal Theology 102, 112, 152
Universality 36, 37, 51, 94
  experience of 156
  of higher values 102, 140
Universe 7,12 45 112, 113, 150
  non-expansion 164
University of Cape Town xvi, 157, 176
  Departments of Mathematics & Applied Mathematics 161
  Summer School 158, 176
University of South Africa 176
University of the Witwatersrand, Johannesburg 177
Unreal, archetypal 140
Unwisdom 90, 109
Upanishads 5, 43, 45, 48, 63, 66, 67, 104, 105, 108, 111, 112, 113, 116, 129, 140, 141
Upliftment 94, 110, 132

Valium, diazepam 57, 58, 88, 89, 91
Valuation 43
Value-building 44
Vedas 112, 135
Vedic 5, 100, 111, 112, 113
Veridicality 25
Verstand 51
Vibration frequencies 22, 174
Viewpoint on free will
  ethical 90
  juridical 90
  scientific 90
  theological 90

Virgin Mother 86
Vishnu 45, 112
Vision of archetypal light 133, 151
Visions 15
Vitzthum, R.C. 2, 50, 149, 150, 151, 162

Wagner, R. 34
Wave functions, equation 12, 168, 169
West, in four-fold cycle 43, 44, 45, 61, 117, 118, 122, 158
Wheel of Life 56, 95
Whiteman J.H.M. xvi, xviii, 175
  archive xx
  biography 175-177
  difficulties with publishing 157
  musical activities 161, 175
  revolutionary thought 2, 152
  teaching and lecturing 157
  website 177
Whiteman, S. xxi, 176, 177
Wicca, School of 59
Wilde, O.F.O.W. 2
Wisdom 75, 76, 77, 78, 90, 96, 109, 110
  source of 96
  ultimate 2, 77
Wisdom traditions 3
Wolf 122
Wolman, B.B. 75
Woman 127
"Word of the Lord" 156
Wordsworth, W. 69, 158
World
  actualisation/manifestation of 31, 167
  ante-world 133, 134
  in the world but not of it 149, 157
  non-physical, other, spiritual 3, 9, 12, 19, 22, 29, 53, 65, 132, 133,140, 160
  parallel 31
  physical 3, 9, 12, 19, 27, 29, 132, 133, 141, 143, 157
  spiritual progress in 144, 145, 148
  projection of 16-fold cycles 42
Worldliness 144
Worlds, spheres
  higher 133, 135, 137
  lower 135, 136
  appearance in 125
World-plan 95, 96
World view 12, 140
Wolman, B.B. 69
Yoga 111
Yoga Sutras 6, 37, 38, 48, 63, 104 - 111, 112, 113, 114, 128, 144
Zion, mystical meaning 142